SpringerBriefs in Mathematical Physics

Volume 2

T0207253

Chiara Esposito

Formality Theory

From Poisson Structures to Deformation
Quantization

 Springer

Chiara Esposito
Department of Mathematics
University of Würzburg
Würzburg
Germany

ISSN 2197-1757   ISSN 2197-1765   (electronic)
ISBN 978-3-319-09289-8 ISBN 978-3-319-09290-4   (eBook)
DOI 10.1007/978-3-319-09290-4

Library of Congress Control Number: 2014945958

Springer Cham Heidelberg New York Dordrecht London

Printed on acid-free paper

Springer is part of Springer Science+Business Media (www.springer.com)

A mio zio Franco

Preface

This book aims to be a survey of the theory of formal deformation quantization of Poisson manifolds, in the formalism developed by Kontsevich. It is intended to be a pedagogical introduction to mathematicians and mathematical physicists who are first approaching the subject. The main topics are the theory of Poisson manifolds, star products and their classification, deformations of associative algebras, and the formality theorem. Furthermore, the aim is to introduce the reader to the relevant physical motivations behind the purely mathematical constructions.

Despite the fact that deformation quantization is a broad research area, the pedagogical literature dealing with this topic, and in particular formal deformation quantization, is limited.

The construction of star products on symplectic manifolds can be found in [3], a monograph focused on Fedosov's approach. Recently, S. Waldmann [6] wrote a more extensive book on the argument, which is only available in German. Further interesting reviews on the subject can be found, for instance, in [1, 2, 4, 5]. In particular, [1] is a sketchy introduction to formal deformation quantization. A detailed, although brief, presentation of the Kontsevich theory from a purely mathematical point of view is included in [2, 4, 5]. It has to be noted that these reviews are oriented to researchers with a basic knowledge of Poisson geometry. There, the authors describe Kontsevich's formality theorem and its relation to the existence and classification of star products on a Poisson manifold. In [4] the author introduces the theory of star products and their classification on symplectic manifolds. The notes in [2, 5] are also focused on further developments of Kontsevich's theory. In particular, the author in [5] also introduces the Tamarkin approach.

The above list, although not exhaustive, represents the state of the art of the introductory literature on formal deformation quantization. At the present time, a book that covers both the basic mathematical tools and Kontsevich's theory, and one that is also designed for first-time readers, is missing. These notes aim to fill this gap writing a didactical exposition to the subject for nonexperts without assuming from the reader too many prerequisites. In particular, this book is addressed to mathematical physicists, who have basic knowledge of differential geometry, classical, and quantum mechanics.

We describe the Poisson structures and their role in classical mechanics, present the results of Kontsevich on the classification of star products on Poisson manifolds, and discuss the physical motivations underlying this theory.

Würzburg, May 2014   Chiara Esposito

References

1. M. Bordemann, in *Deformation Quantization: A Survey*, ed. by J.-C. Wallet. International Conference on Noncommutative Geometry and Physics. J. Phys: Conf. Ser. **103** (2008)
2. A.S. Cattaneo, D. Indelicato, in *Formality and Star Product*, ed. by S. Gutt, J. Rawnsley, D. Sternheimer. Poisson geometry, deformation quantization and group representation. London Math. Soc. Lect. Note Ser. **323** (2004) pp. 81–144
3. B.V. Fedosov, *Deformation Quantization and Index Theory* (Wiley-VCH, 1996)
4. S. Gutt, Deformation quantisation of Poisson manifolds. Geom. Topology Monogr. **17**, pp. 171–220 (2001)
5. B. Keller, *Deformation quantization after Kontsevich and Tamarkin*. In: Déformation, quantification, théorie de Lie, vol. 20 (Panoramas et Synthèses, Société mathématique de France, 2005), pp. 19–62
6. S. Waldmann, *Poisson-Geometrie und Deformationsquantisierung* (Springer-Verlag, 2007)

Acknowledgments

This work is based on the course given during the Summer School of Geometry, which took place in Ouargla (Algeria), 2012. I would like to thank the hospitality and support at the University Kasdi Merbah of Ouargla and in particular Mohamed Amine Bahayou for the nice organization during our stay in Algeria.

I started writing the proposal for this book during my stay in the Oberwolfach Mathematical Institute; I sincerely thank the organization that gave me the possibility of working on this project in such a beautiful and inspiring place.

Special thanks go to Peter Bongaart for his interest in these lectures and for his help during the writing process: without his enthusiasm this book would have never been written. Many thanks also to the Editor of these notes, Aldo Rampioni, for his kind support during all the stages of this project.

I want to thank my advisor, Ryszard Nest, who introduced me to this beautiful field of mathematics and helped me to understand it. Many thanks to Stefan Waldmann for his precious help; his comments have been fundamental to clarify many parts of this book. Thanks also to Thorsten Reichert, for his inspiring seminars on Dolgushev's proof.

I want to thank my friend, Romero Barbieri Solha, for his help in writing the proposal of this book. The Mathematics Department at Würzburg University has been a very pleasant and stimulating working place, and there are several people who contributed to this in particular I would like to thank my dear friends Alejandro Bolaños Rosales, Stephan Hachinger, Filippo Bracci, and Alessandra Zappoli for the help during the last months of writing and editing of this book. Finally, to all the persons who love me and support me, my parents, my brother and my sister, George, Fedele and Susana, Giovannino, my friends from physics, all my family: thank you!

Contents

Chapter 1
Introduction

In these notes we aim to present Kontsevich's formality theorem, which implies both the existence and the classification of star products on Poisson manifolds. This theorem represents a very important breakthrough in the theory of deformation quantization and it has many ramifications.

The philosophy of deformation was proposed by Flato [18] in the seventies and since then, many developments occurred. Deformation quantization is based on such a philosophy in order to provide a mathematical procedure to pass from classical mechanics to quantum mechanics. Quantum mechanics deals with phenomena at nanoscopic scales; these phenomena contradict the laws of classical mechanics and a new fundamental constant enters in the formalism, the Planck constant $\hbar$. The new structures deform, in some sense, the initial ones; in other words, when the new parameter $\hbar$ goes to zero, quantum mechanics coincides with classical mechanics.

The main problem in comparing the classical theory of mechanics and the theory of quantum mechanics is the difference in their mathematical formulation. Indeed, in classical mechanics the observables are functions over the phase space (the flat space $\mathbb{R}^{2n}$ or, more generally, a symplectic or Poisson manifold) while in quantum mechanics the observables are operators in Hilbert spaces of wave functions. This difficulty has been overcome by looking for deformations of algebras of functions over Poisson manifolds: quantum mechanics is then realized in the deformed algebra. The existence of such deformations was proved by Vey [32] in 1975 and few years later the seminal papers [7, 8, 19] in deformation quantization appeared, where the quantization was performed, using the Gerstenhaber's approach [20], by deforming the associative and commutative algebra of classical observables.

More precisely:

Definition 1.1 Let M be a smooth manifold endowed with a Poisson bracket $\{\cdot, \cdot\}$. A star product on M is a "deformation" of the associative algebra of functions $A = C^\infty(M)$ of the form $\star = \sum_{n=0}^\infty t^n P_n$, where the P_n's are bi-differential operators (locally of finite order) such that $P_0(f, g) = f \cdot g$, $P_1(f, g) - P_1(g, f) = 2\{f, g\}$, $f, g \in A$.

C. Esposito, *Formality Theory*, SpringerBriefs in Mathematical Physics,
DOI 10.1007/978-3-319-09290-4_1

The parameter t is taken to be $t = \frac{i\hbar}{2}$ to recover physical results. A star quantization is defined to be a star product on M invariant under some Lie algebra $\mathfrak{g}_0$ of "preferred observables". The invariance property ensures that the classical and quantum evolutions of observables under a Hamiltonian $H \in \mathfrak{g}_0$ coincide [7].

It is worth mentioning that also a spectral theory can be done in this formalism and many results of quantum mechanics can be formulated and solved in this context, for example the spectrum of harmonic oscillators, the hydrogen atom and the angular momentum.

The star product is given by a formal power series and the convergence was studied only in specific examples. There are many interesting mathematical developments, related to the star products. The star representation theory is an important example; the most notable results can be found in [1–3] (nilpotent and solvable Lie groups), [4, 5, 27] (semi-simple Lie groups). Another fundamental application of the star products concerns the theory of quantum groups. This theory is essentially due to Drinfeld [14], who realized that Lie groups and Lie algebras can be deformed by considering their correspondent Hopf algebras. Quantum groups have been largely studied and they have numerous applications in physics (the reader is referred to [15, 25, 29]). Finally, the index theorem [6] has been generalized to deformation quantum algebras [17].

The existence and classification of star products on Poisson manifolds have been major open problems for many years. The regular case was approached with the method of Fedosov [16] and the existence of tangential star-products was established by Masmoudi [26] in 1992. Some concrete examples of star products on non-regular Poisson manifolds appeared already in [7]. The star product on a Lie algebra can be defined starting from the star product on the cotangent bundle of a Lie group (Gutt [21]) and the quantization for some quadratic Poisson structures has been constructed by Omori, Maeda and Yoshika in [28]. Eventually, the existence of star products on any finite-dimensional Poisson manifold was proved by M. Kontsevich as a consequence of his formality theorem [23].

Kontsevich constructed explicitly the map

$$\star : C^\infty(\mathbb{R}^d) \times C^\infty(\mathbb{R}^d) \rightarrow C^\infty(\mathbb{R}^d) : (f, g) \mapsto f \star g \qquad (1.1)$$

and proved that it defines a star product on the Poisson manifold $(\mathbb{R}, \pi)$. Furthermore, he proved that there is a one-to-one correspondence between equivalence classes of star products and equivalence classes of formal Poisson bracket $\pi_t := \sum_{n=0}^{\infty} t^n \pi_n$. This result is a particular consequence of the so-called formality theorem. In order to roughly describe Kontsevich's formality theorem, let M be a smooth manifold and consider the Hochschild complex $C^\bullet(A)$ on the associative algebra $A = C^\infty(M)$ (with pointwise product μ) and its cohomology $HH^\bullet(A)$. They are both differential graded Lie algebras and the formality theorem states that these differential graded Lie algebras are in some sense equivalent. More in detail, denote by $D_{\mathrm{poly}}(M)$ the graded vector space of multidifferential operators $D_{\mathrm{poly}}(M) = \bigoplus_{n=1}^{\infty} D_{\mathrm{poly}}^n(M)$, where $D_{\mathrm{poly}}^n(M) = C^{k+1}(A)$. This graded space can be endowed with a DGLA structure.

Similarly, the graded space of multivector fields on M can be endowed with a DGLA structure. Denote by $T_{\text{poly}}(M) = \bigoplus_{n=1}^{\infty} T_{\text{poly}}^n(M)$, where $T_{\text{poly}}^n(M) = \Gamma(\wedge^{k+1} TM)$ i.e. the space of $k+1$-multivector fields on M. The two DGLAs $T_{\text{poly}}(M)$ and $D_{\text{poly}}(M)$ are quasi-isomorphic complexes [22], which means that the cohomology $HH^\bullet(A)$ of the complex $D_{\text{poly}}(M)$ coincides with the cohomology of $T_{\text{poly}}(M)$ on M. This quasi-isomorphism, unfortunately, does not preserve the Lie bracket, thus it is not a DGLA homomorphism. Kontsevich interpreted the DGLAs $T_{\text{poly}}(M)$ and $D_{\text{poly}}(M)$ in terms of a very general category of objects, called L_∞-algebras. This allowed him to prove, in the formality theorem, the correspondence of such DGLAs.

The main part of the proof provided by Kontsevich consists in the explicit construction of such a correspondence for the local case $M = \mathbb{R}^d$ (the so-called Kontsevich's formula). The physical interpretation of this formula, already suggested by Kontsevich in [23], has been studied by Cattaneo and Felder in [9]. They constructed a topological field theory on a disc and Kontsevich's formula appears as the perturbation series of such a theory, after a suitable renormalization. The method used by Cattaneo and Felder also provides a very nice proof of the existence of Kontsevich's star product [10], which is similar, in the spirit, to Fedosov's construction. More recently, another proof of the global formality was provided by Dolgushev [11].

The formality theorem has ramifications and developments in many directions. First, it is important to mention the operadic approach introduced by Tamarkin [30] in 1998. He observed that for any algebra A its Hochschild complex $C^\bullet(A)$ and its Hochschild cohomology $HH^\bullet(A)$ are algebras over the same operad. This approach will be not treated in these notes (a nice introduction on operads and some important results related to the Tamarkin's approach can be found in [12]) but we want to remark that, using this approach, a new derivation of the formality theorem was found (see [31] for further developments). Kontsevich studied and generalized the Tamarkin's approach in [24], which also contains a systematic study about the generalization of deformation quantization to the case of algebraic varieties. Further developments in the algebro-geometric setting can be found, e.g., in [13].

This book is organized as follows:

Chapter 2 contains a short discussion on the Hamiltonian formulation of classical mechanics. We introduce the description of a classical mechanical system in terms of Poisson manifolds and we define the notion of formal Poisson structures.

Chapter 3 is devoted to the theory of (formal) deformation quantization. We discuss the notion of star product, starting from physical motivations and heading towards its formulation in terms of DGLA. We give a basic introduction of the general theory of deformations via DGLA's, focusing in particular on the examples of multivector fields and multidifferential operators. We introduce Kontsevich's formality theorem, presented as an extension of the Hochschild-Kostant-Rosenberg theorem, and we present some basic tools (e.g. L_∞-algebras, L_∞-morphisms).

In Chap. 4 we aim to give a sketchy exposition of the Kontsevich's formula on $\mathbb{R}^d$. Here we only discuss the globalization approaches of Cattaneo-Felder-Tomassini and Dolgushev, as the Kontsevich proof is extremely technical. Finally we present some open problems related to formal deformation quantization.

A short survey of the notions used throughout this book is given in Appendix A; in particular, we recall the concepts of vector bundles, tensors and connections and we introduce the notions of complexes and cohomologies.

References

1. D. Arnal, J.C. Cortet, Nilpotent Fourier transform and applications. Lett. Math. Phys. **9**, 25–34 (1985)
2. D. Arnal, J.C. Cortet, Star-products in the method of orbits for nilpotent Lie groups. J. Geom. Phys. **2**, 83–116 (1985)
3. D. Arnal, J.C. Cortet, J. Ludwig, Moyal product and representations of solvable Lie groups. J. Funct. Anal. **133**, 402–424 (1995)
4. D. Arnal, M. Cahen, S. Gutt, Representation of compact Lie groups and quantization by deformation. Bull. Acad. Royale Belg. **74**, 123–141 (1988)
5. D. Arnal, M. Cahen, S. Gutt, Star exponential and holomorphic discrete series. Bull. Soc. Math. Belg. **41**, 207–227 (1989)
6. M.F. Atiyah, I.M. Singer, The index of elliptic operators on compact manifolds. Bull. Am. Math. Soc. **69**, 422–433 (1963)
7. F. Bayen, M. Flato, C. Fronsdal, A. Lichnerowicz, D. Sternheimer, Quantum mechanics as a deformation of classical mechanics. Lett. Math. Phys. **1**, 521–530 (1977)
8. F. Bayen, M. Flato, C. Fronsdal, A. Lichnerowicz, D. Sternheimer, Deformation theory and Quantization I-II. Ann. Phys. **111**(61–110), 111–151 (1978)
9. A.S. Cattaneo, G. Felder, A path integral approach to the Kontsevich quantization formula. Comm. Math. Phys. **212**, 591–611 (2000)
10. A.S. Cattaneo, G. Felder, On the globalization of Kontsevich's star product and the perturbative Poisson sigma model. Prog. Theor. Phys. Suppl. **144**, 38–53 (2001)
11. V.A. Dolgushev, Covariant and equivariant formality theorems. Adv. Math. **191**(1), 147–177 (2005)
12. V.A. Dolgushev, C.L. Rogers, Notes on algebraic operads, graph complexes, and Willwacher's construction. Contemp. Math. **583**, 25–147 (2012)
13. V.A. Dolgushev, C.L. Rogers, T. Willwacher, Kontsevich's graph complex, GRT, and the deformation complex of the sheaf of polyvector fields. arXiv:1211.4230 (2012)
14. V. Drinfeld, Quantum groups, in *Proceedings of ICM86, Berkeley*. American Mathematical Society, Providence (1987)
15. P. Etingov, O. Schiffmann, *Lectures on Quantum groups*, Lectures in Mathematical Physics (International Press, Boston, 1998)
16. B.V. Fedosov, A simple geometrical construction of deformation quantization. J. Diff. Geom. **40**, 213–238 (1994)
17. B.V. Fedosov, *Deformation Quantization and Index Theory* (Wiley-VCH, Berlin, 1996)
18. M. Flato, Deformation view of physical theories. Czechoslovak J. Phys. **B32**, 472–475 (1982)
19. M. Flato, A. Lichnerowicz, D. Sternheimer, Crochets de Moyal-Vey et quantification. C. R. Acad. Sci. Paris Sér. A **283**, 19–24 (1976)
20. M. Gerstenhaber, On the Deformation of rings and algebras. Ann. Math. **79**(1), 59–103 (1964)
21. S. Gutt, An explicit ⋆-product on the cotangent bundle of a Lie group. Lett. Math. Phys. **7**, 249–258 (1983)
22. G. Hochschild, B. Kostant, A. Rosenberg, Differential forms on regular affine algebras. Trans. Am. Math. Soc. **102**(3), 383–408 (1962)
23. M. Kontsevich, in *Formality Conjecture*, ed. by D. Sternheimer, et al. Deformation Theory and Symplectic Geometry (Kluwer, Dordrecht, 1997)
24. M. Kontsevich, Deformation quantization of algebraic varieties. Lett. Math. Phys. **56**, 271–294 (2001)

25. S. Majid, *Foundations of Quantum Group Theory* (Cambridge University Press, Cambridge, 1995)
26. M. Masmoudi, Tangential formal deformations of the Poisson bracket and tangential star products on a regular Poisson manifold. J. Geom. Phys. **9**, 155–171 (1992)
27. C. Moreno, Invariant star products and representations of compact semi-simple Lie groups. Lett. Math. Phys. **12**, 217–229 (1986)
28. Y. Omori, H. Maeda, A. Yoshida, Deformation quantizations of Poisson algebras. Contemp. Math. **179**, 213–240 (1994)
29. S. Shnider, S. Sternberg, *Quantum Groups, Volume II of Graduate Texts in Mathematical Physics* (International Press, Boston, 1993)
30. D.E. Tamarkin, Formality of chain operad of little discs. Lett. Math. Phys. **66**, 65–72 (2003)
31. D.E. Tamarkin, B. Tsygan, Cyclic formality and index theorems. Lett. Math. Phys. **56**, 85–97 (2001)
32. J. Vey, Déformation du crochet de Poisson sur une variété symplectique. Commentarii Mathematici Helvetici **50**(1), 421–454 (1975)

Chapter 2
Classical Mechanics and Poisson Structures

In this chapter, we will briefly recall the Hamiltonian formulation of classical mechanics, focusing in particular on its algebraic aspects. In this framework, a classical system will be described by a commutative algebra of functions (classical observables) with the Poisson bracket as a Lie bracket.

We will discuss in detail the properties of the Poisson bracket and introduce the tensor formulation of Poisson structures on manifolds, as the Poisson bracket plays a fundamental role in classical mechanics and in deformation quantization. We will mainly focus on the algebraic rather than geometrical properties of Poisson manifolds, the latter being less important for the theory of deformation quantization. Furthermore, we will introduce the reader to the basic notions needed for the formulation of the formality theory, i.e. formal power series, formal Poisson structures and equivalence classes of formal Poisson structures.

2.1 Hamiltonian Mechanics and Poisson Brackets

This section aims to give a brief introduction to classical mechanics, starting with Newton's laws and heading towards the Hamiltonian approach, with a particular attention to the role of the Poisson bracket. The interested reader is referred to the classical literature on the subject, as e.g. [1, 2, 6] for an exhaustive treatment.

We start by discussing the motion of a point particle of mass m in the Euclidean space $\mathbb{R}^n$. The position of the particle is described by the vector $q := (q^1, \ldots, q^n) \in \mathbb{R}^n$. The vector q is generally parametrized by the variable $t \in \mathbb{R}$. We say that $q(t)$ is the position of the particle at time t.

The velocity $v(t)$ of the particle at time t is defined as

$$v(t) := \dot{q}(t) = \frac{\mathrm{d}q}{\mathrm{d}t}(t), \qquad (2.1)$$

© The Author(s) 2015
C. Esposito, *Formality Theory*, SpringerBriefs in Mathematical Physics,
DOI 10.1007/978-3-319-09290-4_2

where we used Newton's notation $\dot{q}(t)$ to denote the total derivative w.r.t. the time. Similarly, the acceleration is defined as

$$a(t) := \ddot{q}(t) = \frac{d^2 q}{d t^2}(t).$$
(2.2)

The evolution in time of the particle position is described by the n functions $q^i(t)$, $i = 1, \ldots, n$, which are solutions of the set of Newton's equations

$$m\ddot{q}(t) = F(q^1, \ldots q^n),$$
(2.3)

where $F := (F_1, \ldots, F_n)$ denotes the force acting on the particle, together with the initial conditions

$$q(0) = q_0,$$
$$v(0) = v_0.$$
(2.4)

From here on, we assume that the force is conservative, i.e. it can be written in terms of the gradient of a some function $V : \mathbb{R}^n \to \mathbb{R}$

$$F_i = -\frac{\partial V}{\partial q^i}, \qquad i = 1, \ldots, n.$$
(2.5)

The function V is generally called potential.

As will be seen in the following, in the Hamiltonian formalism, the system of second order differential equations (2.3), in the n variables $(q^1, \ldots, q^n)$, becomes a first order system in the $2n$ variables $(q, p) := (q^1, \ldots, q^n, p_1, \ldots, p_n)$. The variables p_i, are called conjugated momenta and they are defined as

$$p_i = m\dot{q}^i.$$
(2.6)

Indeed, using this definition, Newton's equations (2.3) can be rewritten as

$$\dot{q}^i(t) = \frac{p_i(t)}{m},$$
$$\dot{p}_i(t) = -\frac{\partial V}{\partial q^i}(q(t)),$$
(2.7)

and the initial conditions (2.4) read

$$q(0) = q_0,$$
$$p(0) = p_0.$$
(2.8)

Introducing the Hamiltonian function $H : \mathbb{R}^n \times \mathbb{R}^n \to \mathbb{R}^n$

$$H(q, p) := \sum_{i=1}^{n} \frac{p_i^2}{2m} + V(q), \tag{2.9}$$

which represents the energy of the system as a function of the position q and the momentum p, the set of equations (2.7) can be rewritten as the well-known Hamilton's equations:

$$\dot{q}^i(t) = \frac{\partial H}{\partial p_i}(q, p),$$

$$\dot{p}_i(t) = -\frac{\partial H}{\partial q^i}(q, p). \tag{2.10}$$

The $2n$-dimensional space of all the possible positions q and momentum p of a single particle is called phase space and coincides with $\mathbb{R}^n \times \mathbb{R}^n \cong \mathbb{R}^{2n}$. A real-valued smooth function f on the phase space, i.e. $f : \mathbb{R}^{2n} \to \mathbb{R}$, is called classical observable.

Given a generic observable f, it is natural to ask how it evolves in time. Denoting by

$$f_t(q, p) = f(q(t), p(t)), \tag{2.11}$$

the value of the observable at the generic time t, where $q(t)$ and $p(t)$ are solutions of (2.10) with $q(0) = q_0, p(0) = p_0$, we have that

$$\begin{aligned}
\frac{df_t}{dt}(q, p) &= \frac{d}{dt} f(q(t), p(t)) \\
&= \frac{\partial f_t}{\partial q^i} \frac{dq^i}{dt} + \frac{\partial f_t}{\partial p_i} \frac{dp_i}{dt} \\
&= \frac{\partial f_t}{\partial q^i} \frac{\partial H}{\partial p_i} - \frac{\partial f_t}{\partial p_i} \frac{\partial H}{\partial q^i}.
\end{aligned} \tag{2.12}$$

In the above expression, we used the Einstein notation on the sum over repeated indices. We will use this convention throughout these notes. The expression obtained above in (2.12) can be written in a more convenient manner, by introducing the canonical Poisson bracket. This is defined, for two arbitrary functions f and g on the phase space $\mathbb{R}^{2n}$, as

$$\{f, g\} := \frac{\partial f}{\partial q^i} \frac{\partial g}{\partial p_i} - \frac{\partial f}{\partial p_i} \frac{\partial g}{\partial q^i}. \tag{2.13}$$

In this notation, Hamilton's equations read

$$\frac{df_t}{dt} = \{H, f_t\}. \tag{2.14}$$

A given observable f, constant along all solutions $(q(t), p(t))$ of Hamilton's equations (2.10), i.e. $f(q(t), p(t)) = f(q_0, p_0)$ for any $t \in \mathbb{R}$, where $q_0 = q(0)$ and $p_0 = p(0)$, is called (mostly in physics) a constant of motion. It is clear from (2.14) that the Hamiltonian H is always a constant of motion (conservation of the energy of the system).

The above discussion can be generalized to the case in which the phase space is a generic smooth manifold. As will be seen, a classical physical system can be described by the algebra of functions on a given phase space endowed with a Poisson bracket. Because of their importance in the formulation of both quantum and classical mechanics, the Poisson structures will be the main topic in the rest of this chapter.

2.2 Poisson Manifolds

Let M be a smooth manifold. The set $C^\infty(M)$ of real-valued smooth functions on M describes the set of observables. It is a commutative algebra with addition, scalar multiplication and pointwise multiplication given by

$$\begin{aligned}
(\alpha f)(x) &= \alpha f(x), \\
(f + g)(x) &= f(x) + g(x), \\
(f \cdot g)(x) &= f(x)g(x),
\end{aligned} \tag{2.15}$$

for any $f, g \in C^\infty(M)$, $\alpha \in \mathbb{R}$ and $x \in M$.

A Poisson bracket can be defined on $C^\infty(M)$ as follows

Definition 2.1 The bracket operation denoted by

$$\{\cdot, \cdot\} : C^\infty(M) \times C^\infty(M) \to C^\infty(M) \tag{2.16}$$

is called Poisson bracket if it satisfies the following properties

1. $\{f, g\}$ is bilinear with respect to f and g
2. $\{f, g\} = -\{g, f\}$ (skew-symmetry)
3. $\{h, fg\} = f\{h, g\} + \{h, f\}g$ (Leibniz rule)
4. $\{f, \{g, h\}\} + \{g, \{h, f\}\} + \{h, \{f, g\}\} = 0$ (Jacobi identity)

for any $f, g, h \in C^\infty(M)$.

The Poisson bracket makes $C^\infty(M)$ into a Lie algebra, as it satisfies bilinearity, skew-symmetry and the Jacobi identity. It follows that the algebra of observables

$C^\infty(M)$ is a commutative algebra with Poisson bracket as Lie bracket. In other words, it is a Poisson algebra, as defined below.

Definition 2.2 A Poisson algebra is a commutative algebra A with a bracket $\{\cdot, \cdot\}$ making it into a Lie algebra such that it satisfies the Leibniz rule.

Starting from this definition we can introduce the concept of Poisson manifolds in a very natural way as follows

Definition 2.3 A Poisson manifold is a smooth manifold M equipped with a bracket $\{\cdot, \cdot\}$ on its function space $C^\infty(M)$, such that the pair $(C^\infty(M), \{\cdot, \cdot\})$ is a Poisson algebra.

The reader can find a more detailed discussion on Poisson manifolds and their geometrical properties e.g. in [3, 8, 9, 11].

A Poisson manifold can be redefined, in a more modern way, in terms of bivector fields. This formulation is necessary for the description of Kontsevich's theory of deformation quantization of Poisson manifolds. In order to rewrite the definition of Poisson manifolds we need to take a step back.

Let us consider a generic bracket $\{\cdot, \cdot\}$ on $C^\infty(M)$ satisfying the conditions (1)–(3) of Definition 2.1, i.e. bilinearity, skew-symmetry and Leibniz rule. The Leibniz rule implies that, for a given function f on $C^\infty(M)$, the map $g \mapsto \{f, g\}$ is a derivation. Thus, there is a unique vector field X_f on M, called Hamiltonian vector field, such that for any $g \in C^\infty(M)$ we have

$$X_f(g) = \{f, g\}. \tag{2.17}$$

Here

$$X_f(g) = \langle dg, X_f \rangle, \tag{2.18}$$

where dg is the differential of the function $g \in C^\infty(M)$ and $\langle \cdot, \cdot \rangle$ is the pairing between one-forms and vector fields.

In the following, we will express Poisson structures in terms of bivector fields satisfying certain conditions. Recall that $\wedge^2 TM$ is the space of bivector of M: it is a vector bundle over M. A (smooth) bivector field π on M is, by definition, a smooth section of $\wedge^2 TM$, i.e. a map $\pi : M \to \wedge^2 TM$, which associates to each point $m \in M$ a bivector $\pi(m) \in \wedge^2 T_m M$. We denote by $\Gamma(\wedge^2 TM)$ the space of sections on $\wedge^2 TM$.

Given a bivector field π, one can define a bracket $\{\cdot, \cdot\}$ on $C^\infty(M)$ as

$$\{f, g\} := \pi(df, dg) = \langle df \otimes dg, \pi \rangle, \tag{2.19}$$

which satisfies the conditions (1)–(3) of Definition 2.1. It is important to remark that, at this stage, this is not a Poisson bracket, because the Jacobi rule is not a priori satisfied. We sketch the conditions which guarantee this bracket to be a Poisson bracket. A bivector field π such that the bracket defined in Eq. (2.19) satisfies the Jacobi identity is called Poisson tensor or Poisson bivector field. In a local system of coordinates $(x_1, \ldots, x_n)$, Eq. (2.19) can be expressed as

$$\{f, g\} = \pi^{ij} \frac{\partial f}{\partial x^i} \frac{\partial g}{\partial x^j}, \tag{2.20}$$

where π^{ij} are smooth functions on the local chart and are defined by

$$\pi^{ij} = \{x^i, x^j\} = -\pi^{ji}. \tag{2.21}$$

This implies that the bivector field π is locally given by

$$\pi = \frac{1}{2} \pi^{ij} \frac{\partial}{\partial x^i} \wedge \frac{\partial}{\partial x^j}; \tag{2.22}$$

using the local expression (2.20) for the Poisson bracket, we can easily compute the terms of the Jacobi identity:

$$\begin{aligned}
\{\{f, g\}, h\} &= \pi^{ij} \frac{\partial}{\partial x^i} \left(\pi^{kl} \frac{\partial f}{\partial x^k} \frac{\partial g}{\partial x^l} \right) \frac{\partial h}{\partial x^j} \\
&= \pi^{ij} \pi^{kl} \left(\frac{\partial f}{\partial x^i \partial x^k} \frac{\partial g}{\partial x^l} + \frac{\partial f}{\partial x^k} \frac{\partial g}{\partial x^i \partial x^l} \right) \frac{\partial h}{\partial x^j} \\
&\quad + \pi^{ij} \frac{\partial \pi^{kl}}{\partial x^i} \frac{\partial f}{\partial x^k} \frac{\partial g}{\partial x^l} \frac{\partial h}{\partial x^j}.
\end{aligned} \tag{2.23}$$

Similarly we get $\{\{g, h\}, f\}$ and $\{\{h, f\}, g\}$. From the skew-symmetry (2.21) follows that the expressions without derivatives of π^{ij} are invariant under switching $\{ij\} \leftrightarrow \{kl\}$, thus the sum of those three terms yields zero. For this reason, the Jacobi identity reads

$$\pi^{hi} \frac{\partial}{\partial x_h} \pi^{jk} + \pi^{hj} \frac{\partial}{\partial x_h} \pi^{ki} + \pi^{hk} \frac{\partial}{\partial x_h} \pi^{ij} = 0. \tag{2.24}$$

In other words,

Proposition 2.1 *The bivector field $\pi \in \wedge^2 TM$ defines a Poisson bracket in the local coordinates $\{x_i\}_{i=1}^n$ if and only if it satisfies the condition (2.24).*

This condition can be rephrased in an invariant formalism, by introducing the Schouten-Nijenhuis bracket of π. We briefly recall the notion of for a generic multivector field and we prove that π is a Poisson tensor if and only if the Schouten-Nijenhuis bracket $[\pi, \pi]_S$ is vanishing.

The definition of bivector field can be immediately generalized as follows. A k-th multivector field X on a smooth manifold M is a section of the k-th exterior power $\wedge^k TM$ of the tangent bundle TM. In local coordinates $\{x_i\}_{i=1}^n$, the multivector field $X \in \Gamma(\wedge^k TM)$ can be written as

$$X = \sum_{i_1 \dots i_k = 1}^n X^{i_1 \dots i_k}(x) \frac{\partial}{\partial x_{i_1}} \wedge \cdots \wedge \frac{\partial}{\partial x_{i_k}}, \tag{2.25}$$

where the coefficients $X^{i_1 \cdots i_k}(x)$ are smooth functions on M. We denote by $\mathfrak{X}^k(M)$ the space of sections $\Gamma(\wedge^k TM)$; notice that $\mathfrak{X}^0(M) = C^\infty(M)$. It is well-known that, for any vector field $X \in \mathfrak{X}^1(M)$, there is a well defined Lie bracket on vector fields given in terms of Lie derivative $\mathscr{L}_X$:

$$[X, Y] := \mathscr{L}_X Y \quad \forall Y \in \mathfrak{X}^1(M). \tag{2.26}$$

This definition can be also applied to the case in which the second argument is a function:

$$[X, f] := \mathscr{L}_X f = \sum_{i=1}^{n} X^i \frac{\partial f}{\partial x_i}. \tag{2.27}$$

We can extend this bracket to an operation

$$[\cdot, \cdot]_S : \mathfrak{X}^k(M) \otimes \mathfrak{X}^l(M) \to \mathfrak{X}^{k+l-1}(M) \tag{2.28}$$

defined by

$$[X_1 \wedge \cdots \wedge X_k, Y_1 \wedge \cdots \wedge Y_l]_S :=$$

$$\sum_{i=1}^{k} \sum_{j=1}^{l} (-1)^{i+j} [X_i, Y_j] \wedge X_1 \wedge \cdots \wedge \widehat{X}_i \wedge \cdots \wedge X_k \wedge Y_1 \wedge \cdots \wedge \widehat{Y}_j \wedge \cdots \wedge Y_l, \tag{2.29}$$

where the hat denotes the absence of the corresponding term.

Proposition 2.2 *The operation $[\cdot, \cdot]_S$ defined in (2.29) is the unique well-defined $\mathbb{R}$-bilinear local type extension of the Lie derivative $\mathscr{L}_X$ and satisfies*

1. $[X, Y]_S = (-1)^{kl}[Y, X]_S$
2. $[X, Y \wedge Z]_S = [X, Y]_S \wedge Z + (-1)^{(k+1)l} Y \wedge [X, Z]_S$
3. $(-1)^{k(m-1)}[X, [Y, Z]_S]_S + (-1)^{l(k-1)}[[Y, Z]_S, X]_S + (-1)^{m(l-1)}[Z, [X, Y]_S]_S$
   $= 0$

for three multivectors X, Y and Z of degree resp. k, l and m.

This operation is called Schouten-Nijenhuis bracket. Notice that, in particular, for any $f \in C^\infty(M)$ and $X \in \mathfrak{X}^k(M)$,

$$[X, f]_S = -\langle df, X \rangle = \sum_{i=1}^{k} (-1)^i \mathscr{L}_{X_i}(f) X_1 \wedge \cdots \wedge \widehat{X}_i \wedge \cdots \wedge X_k. \tag{2.30}$$

In local coordinates, if

$$X = X^{i_1,\dots,i_n} \frac{\partial}{\partial x_{i_1}} \wedge \cdots \wedge \frac{\partial}{\partial x_{i_n}},$$

$$Y = Y^{j_1,\dots,j_m} \frac{\partial}{\partial x_{j_1}} \wedge \cdots \wedge \frac{\partial}{\partial x_{j_m}}, \tag{2.31}$$

the Schouten-Nijenhuis bracket is given by a $n + m - 1$ contravariant tensor field $[X, Y]_S$

$$[X, Y]_S = X^{li_1,\dots,i_{l-1}i_{l+1},\dots,i_n} \frac{\partial Y^{j_1,\dots,j_m}}{\partial x_l} \frac{\partial}{\partial x_{i_1}} \wedge \cdots \wedge \frac{\partial}{\partial x_{i_{l-1}}} \wedge \frac{\partial}{\partial x_{i_{l+1}}} \wedge$$
$$\wedge \frac{\partial}{\partial x_{j_1}} \wedge \cdots \wedge \frac{\partial}{\partial x_{j_m}} (-1)^n Y^{lj_1,\dots,j_{l-1}j_{l+1},\dots,j_m} \frac{\partial X^{i_1,\dots,i_n}}{\partial x_l} \frac{\partial}{\partial x_{i_1}} \wedge$$
$$\wedge \cdots \wedge \frac{\partial}{\partial x_{i_n}} \wedge \frac{\partial}{\partial x_{j_1}} \cdots \wedge \frac{\partial}{\partial x_{j_{l-1}}} \wedge \frac{\partial}{\partial x_{j_{l+1}}} \wedge \cdots \wedge \frac{\partial}{\partial x_{j_m}}, \tag{2.32}$$

or more succinctly

$$[X, Y]_S^{k_2\dots k_{n+m}} = \varepsilon_{i_2\dots i_n j_1\dots j_m}^{k_2\dots k_{n+m}} X^{l(i_2\dots i_n)} \frac{\partial}{\partial x^l} Y^{j_1\dots j_m}$$
$$+ (-1)^n \varepsilon_{i_1\dots i_n j_2\dots j_m}^{k_2\dots k_{n+m}} Y^{l(j_2\dots j_m)} \frac{\partial}{\partial x^l} X^{i_1\dots i_n}. \tag{2.33}$$

Here

$$\varepsilon_{j_1\dots j_{n+m}}^{i_1\dots i_{n+m}} \tag{2.34}$$

is the Kronecker symbol: it is zero if $(i_1 \dots i_{n+m}) \neq (j_1 \dots j_{n+m})$, and is 1 (resp., -1) if $(j_1 \dots j_{n+m})$ is an even (resp., odd) permutation of $(i_1 \dots i_{n+m})$.

Remark 2.1 The Schouten-Nijenhuis bracket is naturally preserved by any diffeomorphism $\phi : M \to N$. Indeed, we recall that

$$\phi_*[X, Y] = [\phi_* X, \phi_* Y] \qquad X, Y \in \mathfrak{X}^1(N) \tag{2.35}$$

where ϕ_* is the pushforward of a diffeomorphism $\phi : M \to N$. It is easy to check, using the definition of the Schouten-Nijenhuis bracket, that this can be extended to

$$\phi_*[X, Y]_S = [\phi_* X, \phi_* Y]_S, \tag{2.36}$$

for any $X, Y \in \mathfrak{X}^k(M)$ and any diffeomorphism ϕ.

The Schouten-Nijenhuis bracket allows us to characterize a Poisson manifold in a very convenient way.

Theorem 2.1 *A bivector field π is a Poisson tensor if and only if the Schouten-Nijenhuis bracket of π with itself vanishes, i.e.*

$$[\pi, \pi]_S = 0. \tag{2.37}$$

It is easy to check, from Eq. (2.33), that in local coordinates

$$[\pi, \pi]_S = \left(\pi^{hi} \frac{\partial}{\partial x_h} \pi^{jk} + \pi^{hj} \frac{\partial}{\partial x_h} \pi^{ki} + \pi^{hk} \frac{\partial}{\partial x_h} \pi^{ij} \right) \frac{\partial}{\partial x_i} \wedge \frac{\partial}{\partial x_j} \wedge \frac{\partial}{\partial x_k}. \tag{2.38}$$

Then Eq. (2.37) is equivalent to the Jacobi rule (2.24).

Example 2.1 A canonical example is given by $M = \mathbb{R}^{2n}$, with coordinates (q^i, p_i), $i = 1, \ldots, n$. The canonical Poisson bracket of functions on the phase space is defined in Eq. (2.13) and the corresponding Poisson bivector field is

$$\pi = \frac{\partial}{\partial p_i} \wedge \frac{\partial}{\partial q^i}. \tag{2.39}$$

It is easy to check that the bivector π defined above satisfies Eq. (2.37).

Using this characterization of Poisson manifolds and recalling Remark 2.1, we can say that the set of Poisson structures is acted upon by the group of diffeomorphisms on M, that is

$$\pi_\phi := \phi_* \pi, \tag{2.40}$$

where ϕ_* is the pushforward of $\phi : M \to M$. Indeed, by Eq. (2.36) we have

$$[\pi_\phi, \pi_\phi]_S = [\phi_* \pi, \phi_* \pi]_S = \phi_*[\pi, \pi]_S = 0. \tag{2.41}$$

This implies that the set of diffeomorphisms $\phi : M \to M$ defines a gauge group on the set of Poisson structures.

2.3 Formal Poisson Structures

We introduced the Poisson structure on a smooth manifold as a skew-symmetric contravariant bi-tensor which satisfies the Jacobi identity. This structure and its gauge group can be easily extended to formal power series. For this purpose, we briefly recall basic notions and properties of the theory of formal power series.

2.3.1 Formal Power Series

Formal power series are a generalization of power series as formal objects, performed by substituting variables with formal indeterminates. Formal essentially means that there is not necessarily a notion of convergence; formal power series are purely algebraic objects and we essentially use them to represent the whole collection of their coefficients. A detailed discussion on formal power series can be found in [7, 10].

Given a sequence $\{a_n\}_{n \in \mathbb{N}_0}$ of elements on a commutative ring k, a formal power series a is defined by

$$a = \sum_{n=0}^{\infty} t^n a_n \tag{2.42}$$

where t is a formal indeterminate. Two formal power series are equal if and only if their coefficients sequences are the same.

The set of formal power series in t with coefficient in a commutative ring k has also a structure of ring, denoted by $k[\![t]\!]$. Indeed, given two formal power series $a, b \in k[\![t]\!]$, one defines addition of such sequences by

$$a + b = \sum_{n=0}^{\infty} t^n (a_n + b_n), \qquad a_n \in k, \tag{2.43}$$

and multiplication by

$$ab = \sum_{n=0}^{\infty} t^n c_n, \qquad c_n = \sum_{k=0}^{n} a_k b_{n-k}, \qquad a_n, b_n \in k. \tag{2.44}$$

With these two operations, the set $k[\![t]\!]$ becomes a commutative ring with 0 element $(0, 0, \dots)$, multiplicative identity $(1, 0, 0, \dots)$ and the invertible elements are the series with non-vanishing constant term.

Given a vector space V over the ring k, we denote by $V[\![t]\!]$ the space of formal power series with coefficients in V,

$$v = \sum_{n=0}^{\infty} t^n v_n, \qquad v_n \in V. \tag{2.45}$$

Elements in $V[\![t]\!]$ can also be summed term by term and

$$av = \sum_{n=0}^{\infty} t^n c_n, \qquad c_n = \sum_{k=0}^{n} a_k v_{n-k}, \qquad a \in k[\![t]\!], \ v \in V[\![t]\!]. \tag{2.46}$$

In other words, $V[\![t]\!]$ becomes a $k[\![t]\!]$-module. The order of a formal power series is defined by the minimum of the set of all non-negative integers n such that $a_n \neq 0$ and is denoted by $o(v)$. If $v = 0$ the order is defined to be $+\infty$. Furthermore, $V[\![t]\!]$ can be endowed with a metric defined by

$$d : V[\![t]\!] \times V[\![t]\!] \to \mathbb{R} : (v, w) \mapsto d(v, w) := \begin{cases} 2^{-o(v-w)}, & \text{if } v \neq w \\ 0, & \text{if } v = w. \end{cases} \tag{2.47}$$

It induces a Hausdorff topology, called the t-adic topology on $V[\![t]\!]$.

Lemma 2.1 *Let V_1 and V_2 be two k-modules and $\Phi : V_1[\![t]\!] \to V_2[\![t]\!]$ be a $k[\![t]\!]$-linear map. Then, for any non-negative integer r there is a unique linear map $\Phi_r : V_1 \to V_2$ such that*

$$\Phi(v) = \sum_{r=0}^{\infty} t^r \sum_{s=0}^{r} \Phi_s(v_{r-s}) \tag{2.48}$$

for all $v = \sum_{r=0}^{\infty} t^r v_r \in V_1[\![t]\!]$.

If k is a commutative ring, this Lemma can be generalized to the case of k-multilinear maps.

It is important to remark that if A is an algebra over the commutative ring k, the set of formal power series $A[\![t]\!]$ with coefficients in A

$$a = \sum_{n=0}^{\infty} t^n a_n, \quad a_n \in A \tag{2.49}$$

forms an algebra over the ring $k[\![t]\!]$. In fact, elements in $A[\![t]\!]$ can be composed by

$$ab = \sum_{n=0}^{\infty} t^n c_n \quad c_n = \sum_{k=0}^{n} a_k b_{n-k}, \quad a_n, b_n \in A, \tag{2.50}$$

as $A[\![t]\!]$ is a $k[\![t]\!]$-module. Notice that $A[\![t]\!]$ is an algebra of the same type of A; in particular if A is unital associative, $A[\![t]\!]$ will be also unital and associative.

Let U be an open set in $\mathbb{R}^n$ such that $0 \in U$ and let $f \in C^{\infty}(U)$. We denote by $\widehat{f}$ the formal power series

$$\widehat{f} = \sum_{n=0}^{\infty} \frac{t^n}{n!} f^{(n)}(0) \tag{2.51}$$

where $f^{(n)}$ is the n-th derivative of the function f. A fundamental property of formal power series is given by the following

Theorem 2.2 (Borel Lemma, first version) *Given a sequence of real numbers* $\{a_n\}$ *of non-negative integers, there exists a smooth function* $f \in C^\infty(U)$ *such that*

$$\frac{1}{n!}f^{(n)}(0) = a_n \in \mathbb{C}[\![t]\!]. \tag{2.52}$$

In other words, the mapping from $C^\infty(U)$ *to the ring of formal power series* $C^\infty(\mathbb{R})[\![t]\!]$ *given by* $f \mapsto \hat{f}$ *is a* $\mathbb{R}$*-linear surjective algebra homomorphism.*

The surjectivity of the map defined by is quite hard to prove; on the other hand, the linearity is evident and we have

$$(fg)^{(n)}(0) = \sum_{s=0}^{r}\binom{r}{s}f^{(s)}(0)g^{(r-s)}(0), \tag{2.53}$$

which implies $\widehat{fg} = \hat{f}\hat{g}$. An elementary proof of Borel's lemma can be found in [4] and in [5]. This lemma implies that we can view the formal power series as the (formal) Taylor expansion of a smooth function at zero.

2.3.2 Formal Poisson Structures

In order to extend Poisson structures to formal power series, we need to figure out what a formal multivector field is. Using the notion of formal power series with coefficients on a vector space discussed above, we can introduce the concept of formal vector field as follows.

Definition 2.4 A formal vector field is a formal power series

$$X = \sum_{n=0}^{\infty}X_n t^n, \quad X_n \in \mathfrak{X}^1(M). \tag{2.54}$$

The set of formal vector fields is denoted by $\mathfrak{X}^1(M)[\![t]\!]$. This definition can be immediately extended to multivector fields, thus a formal multivector field is an element in $\mathfrak{X}^k(M)[\![t]\!]$, i.e. a formal power series with coefficients in $\mathfrak{X}^k(M)$. Finally, we can define the extension of Poisson structures to formal power series as follows:

Definition 2.5 A formal Poisson structure is a formal power series

$$\pi_t = \pi_0 + t\pi_1 + t^2\pi_2 + \cdots = \sum_{n=0}^{\infty}t^n\pi_n \in \mathfrak{X}^2(M)[\![t]\!], \tag{2.55}$$

where the π_n's are skew-symmetric vector fields on M, such that the Schouten-Nijenhuis bracket of π_t with itself vanishes order by order,

$$[\pi_t, \pi_t]_S = 0. \tag{2.56}$$

Given a Poisson manifold (M, π), the formal Poisson structure can be interpreted as a formal deformation of the structure π by setting $\pi_0 = \pi$; the requirement (2.56) gives k equations, i.e.

$$[\pi, \pi]_S = 0, \qquad \text{order } 0$$
$$[\pi, \pi_1]_S + [\pi_1, \pi]_S = 0, \qquad \text{order } 1 \tag{2.57}$$

and generally, at order $k \geq 2$

$$[\pi, \pi_k]_S = -\frac{1}{2} \sum_{l=1}^{k-1} [\pi_l, \pi_{k-l}]_S. \tag{2.58}$$

A formal Poisson structure on M induces a Lie bracket on $C^\infty(M)[\![t]\!]$ by

$$\{f, g\}_t := \sum_{n=0}^\infty t^n \sum_{\substack{i,j,k=0 \\ i+j+k=n}}^n \pi_i(df_j, dg_k), \tag{2.59}$$

where

$$f = \sum_{j=0}^\infty t^j f_j \qquad \text{and} \qquad g = \sum_{k=0}^\infty t^k g_k. \tag{2.60}$$

We recall that the gauge group on the set of Poisson structures is given by the diffeomorphisms on M and the action is given by

$$\pi_\phi = \phi_* \pi. \tag{2.61}$$

To extend this action to the set of formal Poisson structures we consider *paths* of formal diffeomorphisms of M which start at the identity Id_M diffeomorphism. More explicitly, consider the one-parameter group of diffeomorphisms ϕ_t on M with $\phi_0 = \mathrm{Id}_M$. Given a Poisson structure π on M, ϕ_t defines a *deformed* Poisson structure by

$$\pi_t = (\phi_t)_* \pi. \tag{2.62}$$

Using Eq. (2.51) we can find the formal version of π_t. Since ϕ_t is the flow of a vector field X on M, we have

$$\frac{d(\phi_t)_*}{dt} = \mathcal{L}_X (\phi_t)_* = (\phi_t)_* \mathcal{L}_X, \tag{2.63}$$

for any t. It follows that

$$\left.\frac{d^n}{dt^n}\right|_{t=0} \pi_t = \left.\frac{d^n}{dt^n}\right|_{t=0} (\phi_t)_* \pi = (\mathscr{L}_X)^n \pi. \tag{2.64}$$

Thus, using Eq. (2.51), the formal power series of π_t is given by

$$\widehat{\pi}_t = \sum_{n=0}^{\infty} \left.\frac{d^n}{dt^n}\right|_{t=0} \pi_t = \pi + t\mathscr{L}_X \pi + \frac{t^2}{2}(\mathscr{L}_X)^2 \pi + \cdots =: \exp(t\mathscr{L}_X)\pi. \tag{2.65}$$

In other words, the gauge group is given by the formal diffeomorphism $\phi_t = \exp(t\mathscr{L}_X)$. Notice that the structure of a group is given by the formula (BCH):

$$\exp(tX) \cdot \exp(tY) := \exp\left(tX + tY + \frac{1}{2}t[X, Y] + \cdots\right). \tag{2.66}$$

We can generalize the above discussion to the case in which X is a formal vector field and we can define the formal diffeomorphism on M as a $\mathbb{R}[[t]]$-linear map $\phi_t : \mathfrak{X}^k(M)[[t]] \to \mathfrak{X}^k(M)[[t]]$ of the form $\phi_t = \exp(\mathscr{L}_X)$ with $X \in t\mathfrak{X}^1(M)[[t]]$. Finally we can define the equivalence class of formal Poisson structures as follows

Definition 2.6 Two formal Poisson structures π_t and $\tilde{\pi}_t$ are said to be equivalent if there exists a formal diffeomorphism such that

$$\pi_t = \exp(t\mathscr{L}_X)\tilde{\pi}_t. \tag{2.67}$$

References

1. R. Abraham, J.E. Marsden, *Foundations of Mechanics*, 2nd edn. (AMS Chelsea Publications, New York, 1980)
2. V.I. Arnold, *Mathematical Methods of Classical Mechanics*, 2nd edn. (Springer, New York, 1997)
3. A. Cannas da Silva, A. Weinstein, *Geometric Models for Noncommutative Algebras*, Berkeley Mathematics Lecture Notes series (AMS, Providence, 1999)
4. B. Casselman, Variations on a theorem of Émile Borel. Available on Casselman's webpage http://www.math.ubc.ca/cass/research/pdf/Emile.pdf (2012)
5. L. Hörmander, *The Analysis of Linear Partial Differential Operators I: Distribution Theory and Fourier Analysis* (Springer, Berlin, 1990)
6. J.E. Marsden, *Introduction to Mechanics and Symmetry*, vol. 17 (Springer, New York, 1999)
7. J.M. Ruiz, *The Basic Theory of Power Series*, Advanced Lectures in Mathematics (Vieweg, Braunschweig, 1993)
8. I. Vaisman, *Lectures on the Geometry of Poisson Manifolds* (Birkhäuser, Berlin, 1994)
9. S. Waldmann, *Poisson-Geometrie und Deformationsquantisierung: Eine Einführung* (Springer, Berlin, 2007)
10. H.S. Wilf, *Generatingfunctionology*, 3rd edn. (CRC Press, Natick, 2005)
11. N.T. Zung, J-P. Dufour, *Poisson Structures and Their Normal Forms* (Springer, Berlin, 2005)

Chapter 3
Deformation Quantization and Formality Theory

This chapter will be devoted to the theory of formal deformation quantization. We will recall the description of a quantum physical system in terms of a non-commutative algebra of operators (quantum observables) on a Hilbert space.

The theory of deformation quantization aims to formalize the passage from classical physics to quantum physics using the Dirac quantization rules as guideline. The first requirement for such a theory to be consistent is the existence of a classical limit: a quantum system has to reduce to the corresponding classical system when the limit of $\hbar$, the Planck constant, approaches zero. Therefore, the quantization of a classical system should consist of a deformation of the same system in the parameter $\hbar$. Moreover, the quantization rules state that to any classical observable there corresponds a quantum observable and that the Poisson bracket corresponds to the quantum commutator of corresponding observables. These requirements can be implemented by demanding that the quantization of a classical system shall be given by a star product, an associative non-commutative deformation of the usual product on the algebra of classical phase space functions, that depends on $\hbar$ and such that the associated commutator is a deformation of the Poisson bracket.

We will give a historical overview of the results by Groenewold [28] on the construction of maps between classical and quantum observables, and of the notion of star product on symplectic manifolds, the conditions for its existence [3] and the generalizations provided by De Wilde and Lecomte [16] and Fedosov [20].

We will then define the star product as a formal deformation of an algebra, show its relation with the Poisson structure and, after defining the concept of equivalent star products, we will discuss the problem of classification. The above discussion will make clear that the main mathematical problem of deformation quantization is the construction of a star product. In his well-known paper [38], Kontsevich proved that, as a consequence of the formality theorem, this construction is possible for any Poisson manifold and he solved the classification problem. This result will need the introduction of the theory of deformations of associative algebras by Gerstenhaberand

© The Author(s) 2015
C. Esposito, *Formality Theory*, SpringerBriefs in Mathematical Physics,
DOI 10.1007/978-3-319-09290-4_3

the Hochschild–Kostant–Rosenberg theorem. Eventually, we will state Kontsevich's formality theorem and we will show that it implies that every Poisson manifold admits a formal quantization.

3.1 Quantum Mechanics: Standard Picture

In the previous chapter we gave a brief introduction of classical mechanics, focusing in particular on the Hamiltonian formulation. This picture can not be applied to the description of physical phenomena at microscopic scales, where the action is of the order of the Planck constant $\hbar$. Quantum mechanics was born at the beginning of the 20th century to describe these phenomena. In this section we aim to give a brief introduction of the standard formulation of quantum mechanics; we recall the concepts of states and observables, emphasizing the differences with the classical ones, and we discuss the time evolution of a quantum observable in the Heisenberg formulation. The reader is referred to the classical literature on the subject, as e.g. [18, 42]. A nice presentation of the mathematical foundations of quantum mechanics is given in [8].

In Sect. 2.1 we showed that, in classical mechanics, we can make simultaneous predictions of conjugate variables, by solving Newton's equations (2.7). In quantum mechanics this is not possible and we can only predict the probability of outcomes of concrete experiments; the uncertainty that we have in these predictions is quantified by the Heisenberg principle. For this reason we say that quantum mechanics is a probabilistic theory and it deeply changed the philosophical concept of our knowledge of reality. This forces us to use, in quantum mechanics, a different mathematical formulation from the classical one discussed in Sect. 2.1.

First, we recall that in classical mechanics, the state of a physical system is described by a point in the phase space. In quantum mechanics states are represented by unit vectors in a given (complex separable) Hilbert space $\mathscr{H}$. The physical system determines the nature of such a Hilbert space; for instance, as we will see in the following, the Hilbert space for positions and momentum states is the space of square-integrable functions. An observable, described in classical mechanics by a function on the phase space, is represented by a linear self-adjoint operator $\widehat{f}$ on the Hilbert space $\mathscr{H}$.

The canonical quantization of a classical system on the phase space $\mathbb{R}^{2n}$ can be performed by means of the correspondence principle; as already stated, the Hilbert space associated to such a physical system is given by the square-integrable functions $L^2(\mathbb{R}^n)$ on $\mathbb{R}^n$ and we can associate to the classical observables q^i and p_j the quantum operators $\widehat{q}^i$ and $\widehat{p}_j$ as follows

$$q^i \rightarrow \widehat{q}^i = q^i,$$
$$p_j \rightarrow \widehat{p}_j = -i\hbar \frac{\partial}{\partial q^i}. \tag{3.1}$$

These operators satisfy the canonical commutation relations

$$[\widehat{q}^i, \widehat{p}_j] = i\hbar \delta^i_j,$$

(3.2)

and it is well known that these relations immediately led to the Heisenberg principle. Finally, the Poisson bracket is mapped into the commutator of operators:

$$\{f, g\} \rightarrow -\frac{i}{\hbar}[\widehat{f}, \widehat{g}].$$

(3.3)

This rule is not well-defined, as it shows an ambiguity when products of classical observables are involved; for instance,

$$\{q^3, p^3\} + \frac{1}{12}\{\{p^2, q^3\}, \{x^2, p^3\}\} = 0,$$

$$\frac{1}{i\hbar}\left[\widehat{q}^3, \widehat{p}^3\right] + \frac{1}{12i\hbar}\left[\frac{1}{i\hbar}\left[\widehat{p}^2, \widehat{q}^3\right], \frac{1}{i\hbar}\left[\widehat{q}^2, \widehat{p}^3\right]\right] = -3\hbar^2.$$

(3.4)

The extra term $-3\hbar^2$ is not predicted by application of the canonical quantization rule.

In classical mechanics, given a physical observable f, its time evolution is governed by the equation

$$\frac{df}{dt} = \{H, f\},$$

(3.5)

where H is the Hamiltonian function associated to the physical system. Applying the correspondence principle to this equation, i.e. substituting the operator $\widehat{f}$ to the function f and the commutator of operators to the Poisson bracket we get

$$\frac{d\widehat{f}}{dt} = \frac{i}{\hbar}[\widehat{H}, \widehat{f}],$$

(3.6)

where $\widehat{H}$ is the Hamiltonian operator, the observable associated to the energy of the physical system. The Hamiltonian operator is, as in classical mechanics, a constant of motion. Equation (3.6) coincides with the Heisenberg formulation for the time evolution of a quantum observable.

This short presentation of the quantum framework allows us to focus on an important difference between classical and quantum physical systems. Indeed, on the one hand a classical system is described by the commutative algebra of smooth functions on a manifold; on the other hand a quantum system is described by a non-commutative algebra of operators. This observation led to the idea that quantum mechanics can be regarded as a deformation of classical mechanics. In the following section we discuss the general problem of the quantization and, in particular, the problems related to a precise mathematical formulation of the correspondence between classical and quantum systems.

3.2 Quantization: Ideas and History

The main goal of this section is to introduce the reader the basic ideas of deformation quantization. First, we introduce the fundamental problem of quantization and we briefly review the different approaches that have been developed. Then we describe more carefully the deformation quantization approach and we present the historical developments. Interesting reviews on the subject can be found, for instance, in [10, 14, 29, 36]. Furthermore, Sternheimer [53] describes the birth of deformation quantization and its historical evolution.

In the previous section, we discussed what a physical system is in quantum mechanics and we underlined the differences between the classical and the quantum framework. Quantization is a procedure to pass from classical to quantum mechanics and it is natural to ask whether there is a precise mathematical formulation for such a procedure, which solves the ambiguity of the canonical quantization discussed above. As already pointed out, from Eq. (3.2) we can observe that the observables in quantum mechanics, unlike those in classical mechanics, do not commute with one another. For this reason, the first attempt to quantize a classical system could consist in looking for a correspondence $Q: f \mapsto Q(f)$, mapping a function f to a self-adjoint operator $Q(f)$ on a Hilbert space $\mathscr{H}$, which gives us the non-commutative structure of the algebra of operators from the commutative one $C^\infty(M)$. The correspondence Q should satisfy the following properties

1. $Q(1) = \mathbb{I}$, $Q(q) = \widehat{q}$ and $Q(p) = -i\hbar \frac{\partial}{\partial q}$,
2. $f \mapsto Q(f)$ linear,
3. $[Q(f), Q(g)] = i\hbar Q(\{f, g\})$,
4. for any $\phi: \mathbb{R} \to \mathbb{R}$, $Q(\phi \circ f) = \phi(Q(f))$.

The condition (4) is usually known as the von Neumann rule and it essentially ensures that a polynomial of function gets mapped into a polynomial of operators. Unfortunately, there is no such correspondence, as these properties are mutually inconsistent (the inconsistency appears even if we require only the conditions (1), (2) and (4)) [1]. Indeed Groenewold proved in [28] that the Poisson algebra $C^\infty(M)$ can not be quantized in such a way that the Poisson bracket of two classical observables is mapped into the Lie bracket of the correspondent operators.

There are different mathematical approaches to this quantization problem. A first approach was given by geometric quantization, which aims to find a relation between the phase space and the corresponding Hilbert space. It accepts the conditions (1)–(3) but restricts the space of quantizable observables to exclude problematic terms (as the ones showed in Eq. (3.4)). This approach is due to Souriau [52], Konstant [39] and Segal [51]. The quantization of a particular class of Kähler manifolds was studied in Berezin's quantization [4, 5]. Finally, the approach we are most interested in, the so-called deformation quantization, is a quantization procedure which satisfies properties (1), (2) and the condition (3) only asymptotically in the limit $\hbar \to 0$. This theory was introduced in [22, 23] and developed in [3], where the authors suggest to deform the pointwise product of functions to get a non-commutative one.

This implies that quantization is "a deformation of the structure of the algebra of classical observables rather than a radical change in the nature of the observables". In the following, the non-commutative product (called star product) is given by a formal deformation of the algebraic structure of $C^\infty(M)$. The concept of formal deformation will be clarified in next sections, but here we want to remark that it refers to the fact that the star product is given as a formal power series (defined in Sect. 2.3.1). A non-formal approach to the problem of quantization, called strict deformation quantization, produces quantum algebras of observables not just in the space of formal power series but in terms of C^*-algebras, as suggested by Rieffel [50]. The relation between formal deformation quantization and strict deformation quantization has been subject of several studies. The basic idea is that, given a formal deformation quantization, the subalgebra of converging power series should give somehow a strict deformation quantization, but the only example where this relation is clear is given by the standard Poisson structure on $\mathbb{R}^{2n}$. Convergence of formal power series in formal deformation quantization has been studied by several authors, e.g. [11, 46]. In the present notes we focus our attention on the formal deformation quantization, referring the reader to [48, 49] for a precise presentation of strict deformation quantization.

The origins of the (formal) deformation quantization can be found in Weyl's quantization procedure [57]; given a classical observable $u(p,q)$ on the phase space $\mathbb{R}^n$, Weyl found an explicit formula to associate to u an operator (the quantum observable) $\Omega(u)$ in the Hilbert space $L^2(\mathbb{R}^{2n})$:

$$u \mapsto \Omega(u) := \int_{\mathbb{R}^{2n}} \tilde{u}(\xi, \eta) e^{\frac{i}{\hbar}(\widehat{p}\cdot\xi + \widehat{q}\cdot\eta)} d^n\xi \, d^n\eta, \tag{3.7}$$

where $\tilde{u}$ is the inverse Fourier transform of u and $\widehat{p}_i$ and $\widehat{q}_j$ are operators satisfying the canonical commutation relation (3.2); here the integral is taken in the weak operator topology. Subsequently, Wigner [58] found an inverse formula, which maps an operator into its symbol and Moyal [43] found an explicit formula for the symbol of a quantum commutator (which is now called Moyal bracket):

$$M(u, v) = \frac{\sinh(tP)}{t} = P(u, v) + \sum_{k=1}^{\infty} \frac{t^{2k}}{(2k+1)!} P^{2k+1}(u, v), \tag{3.8}$$

where $t = \frac{i\hbar}{2}$ and P^k is the k-th power of the Poisson bracket (2.13) on $\mathbb{R}^{2n}$. The classical symbol of a product $\Omega(u)\Omega(v)$ had already been found by Groenewold [28] and can be interpreted as the first appearance of the Moyal product (denoted by $\star_M$); indeed we can rewrite the above bracket as

$$M(u, v) = \frac{1}{2t}(u \star_M v - v \star_M u). \tag{3.9}$$

The interpretation of this product as a non-commutative deformation of the pointwise product on the algebra of classical observables is due to Flato et al. and can be found in their seminal paper [3]. They used Gerstenhaber's deformation theory to describe quantum mechanics as a deformation of classical mechanics and found several applications. In particular, they proved the existence of a star product on a generic symplectic manifold (a symplectic manifold is a pair (M, ω), where M is a smooth manifold and ω is a closed, non-degenerate 2-form on M) admitting a flat connection.

The problem of existence of deformation quantization on a generic symplectic manifold had been further developed (some interesting discussions on the topic can be found already in [44, 54]) and was solved by de Wilde and Lecomte [16], using cohomological arguments. Independently, the existence of star products was proved by Omori et al. [47] and a few years earlier by Fedosov [20] (we invite the reader to refer to Fedosov's book [21]).

In subsequent works, the problem of the classification of equivalent star products was also settled by several authors. The equivalence of deformations had already been studied by Gerstenhaber [25] and Flato et al. claimed in [2] that the equivalence is linked with the second de Rham cohomology of any symplectic manifold. This result has been proved with different approaches, first by Nest and Tsygan [45], then Deligne [17] and Bertelson et al. [7].

The existence and classification of star products for any finite dimensional Poisson manifold has been first conjectured [37] and then proved [38] by Kontsevich. Afterwards, parametrizations of equivalence classes of special star products have been obtained, e.g. star products with separation of variables [35], invariant star products on a symplectic manifold with an invariant symplectic connection [6] and algebraic star products [15].

In the following sections we introduce Kontsevich's theory and in the next chapter we briefly discuss Kontsevich's formula of a star product on $\mathbb{R}^n$ and further developments of this theory.

3.3 Star Products and Classification

We introduced the main idea of deformation quantization, the deformation of the classical algebra of functions to get a non-commutative one (the so-called star product), which would give a description of the quantum algebra of observables. In this section we start discussing some more technical aspects; in particular, we define a formal deformation of a generic algebra and we discuss the star product as a particular case of formal deformation. These definitions allow us to understand how crucial the role of the Poisson bracket is in deformation quantization; we introduce the notion of equivalence of star products and formulate the problem of their classification. Furthermore, we state the main result obtained by Kontsevich [38], which solves the classification problem of star products.

Let k be a commutative ring and A an algebra over k. Let us consider the ring $k[[t]]$ of formal power series in t and the algebra $A[[t]]$ of formal power series over $k[[t]]$ with coefficients in A.

Definition 3.1 A formal deformation of the multiplication of A is a $k[[t]]$-bilinear map

$$\star : A[[t]] \times A[[t]] \to A[[t]] \tag{3.10}$$

such that, for any $u, v \in A[[t]]$

$$u \star v = uv \qquad \mod t \tag{3.11}$$

where uv is the multiplication of formal power series defined in Eq. (2.50).

The star product is first defined on A: the product of two elements $a, b \in A$ is given by

$$a \star b = a \cdot b + P_1(a, b)t + P_2(a, b)t^2 + \cdots + P_n(a, b)t^n + \cdots \tag{3.12}$$

where P_i's are k-bilinear maps on A and we denote by $\cdot$ the multiplication of A. Putting $P_0(a, b) = a \cdot b$ we can write

$$a \star b = \sum_{n=0}^{\infty} P_n(a, b)t^n. \tag{3.13}$$

The extension to formal power series by $k[[t]]$-bilinearity is given by

$$\left(\sum_{k=0}^{\infty} f_k t^k \right) \star \left(\sum_{l=0}^{\infty} g_l t^l \right) = \sum_{n=0}^{\infty} \left(\sum_{k+l+m=n} P_m(f_k, g_l) \right) t^n \tag{3.14}$$

for $f_k, g_l \in A$. The $\star$-product is associative if, for any $n \geq 0$

$$\sum_{i+j=n} P_i(P_j(a, b), c) = \sum_{i+j=n} P_i(a, P_j(b, c)), \quad a, b, c \in A \tag{3.15}$$

Again, the associativity can be extended to $A[[t]]$ by $k[[t]]$-bilinearity. It is evident that a necessary condition for the associativity of $\star$ is that the multiplication $a \cdot b$ is associative. We also remark that the star product is a continuous operation with respect to the t-adic topology introduced in Sect. 2.3.1. Assume that A is associative and commutative, with a unit element 1. An associative formal deformation of A is given by a formal power series (3.13), where $\{P_n\}_{n \in \mathbb{N}_0}$ is a sequence of k-bilinear maps such that the product $\star$ is associative. It can be proved that an associative formal deformation of A admits a unit element.

Furthermore, defining

$$\{a, b\} := P_1(a, b) - P_1(b, a), \quad a, b \in A, \tag{3.16}$$

we can easily prove that the operation $\{\cdot, \cdot\}$ is a Poisson bracket on A, as it satisfies the requirements of Definition 2.1. First, we observe that the bracket

$$[a, b]_\star := \frac{1}{t}(a \star b - b \star a) \tag{3.17}$$

is a Lie bracket. Indeed, it is skew-symmetric and bilinear by definition and the Jacobi identity follows from the associativity of the $\star$ product. The bracket $\{\cdot, \cdot\}$ equals the reduction modulo t of $[\cdot, \cdot]_\star$; thus it is still a Lie bracket. Finally, the Leibniz identity of $[\cdot, \cdot]_\star$ is also a consequence of the associativity of $\star$, thus we can conclude that the operation $\{\cdot, \cdot\}$ is a Poisson bracket on A.

An example of particular interest is given by the algebra $C^\infty(M)$ of real-valued smooth functions on a smooth manifold M introduced in Sect. 2.2, where the product is given by

$$f \cdot g(x) := f(x)g(x), \quad \forall x \in M \quad \text{(pointwise product)} \tag{3.18}$$

and is clearly associative and commutative. Recall that an application $P : C^\infty(M) \times \cdots \times C^\infty(M) \to C^\infty(M)$ is a multidifferential operator if in each local coordinates $(x_1, \ldots, x_n)$ on M we have

$$P(f_1, \ldots, f_m) = \alpha_{k_1 \ldots k_n} \frac{\partial f_1}{\partial x^{k_1}} \cdots \frac{\partial f_n}{\partial x^{k_m}}, \tag{3.19}$$

where k_i are multi-indices and $\alpha_{k_1 \ldots k_n}$ are smooth functions locally defined.

Definition 3.2 A star product on M is an associative formal deformation of $C^\infty(M)$

$$\star : C^\infty(M)[[t]] \times C^\infty(M)[[t]] \to C^\infty(M)[[t]], \tag{3.20}$$

given by

$$f \star g = f \cdot g + \sum_{n=1}^{\infty} P_n(f, g)t^n, \tag{3.21}$$

where the $\mathbb{R}$-bilinear maps $P_n : C^\infty(M) \times C^\infty(M) \to C^\infty(M)$ are bi-differential operators.

As pointed out in [14], the P_i's could in principle be just bilinear maps; bidifferential operators are defined only locally, so this requirement encodes the locality of quantum physics. More precisely, one requires that, in local coordinates,

$$P_i(f, g) = \sum_{K,L} \alpha_i^{KL} \frac{\partial f}{\partial x^K} \frac{\partial g}{\partial x^L}, \tag{3.22}$$

where $K = (k_1, \ldots, k_m)$ and $L = (l_1, \ldots, l_n)$. The α_i^{KL}'s are smooth functions, which are non-zero only for finitely many choices of K and L. Furthermore, we require that the unit element of $C^\infty(M)$ is preserved by the star product, i.e.

$$f \star 1 = 1 \star f = f; \tag{3.23}$$

using the expression (3.21) we have

$$P_n(f, 1) = P_n(1, f) = 0 \quad \forall n \geq 1. \tag{3.24}$$

This means that the P_n's are bi-differential operators with no term of order 0.

As discussed above, the operation $\{f, g\} := P_1(f, g) - P_1(g, f)$ is a Poisson bracket for any elements $f, g \in C^\infty(M)$. This implies that M is a Poisson manifold with Poisson structure given by

$$\{f, g\} = \pi(\mathrm{d}f, \mathrm{d}g), \quad \pi \in \mathfrak{X}^2(M). \tag{3.25}$$

Example 3.1 (Moyal Product) The simplest example of a deformed product on $C^\infty(\mathbb{R}^{2n})$ is the Moyal product. Let us consider the manifold $M = \mathbb{R}^{2n}$ with the $2n$-variables

$$(q, p) = (q_1, \ldots, q_n, p_1, \ldots, p_n)$$

We can give an explicit formula for the product of two elements $f, g \in C^\infty(\mathbb{R}^{2n})$

$$f \star g(q, p) = f(q, p) \, \exp\left(\frac{i\hbar}{2}\left(\frac{\overleftarrow{\partial}}{\partial q}\frac{\overrightarrow{\partial}}{\partial p} - \frac{\overleftarrow{\partial}}{\partial p}\frac{\overrightarrow{\partial}}{\partial q}\right)\right) g(q, p), \tag{3.26}$$

where the $\overleftarrow{\partial}$'s operate on f and the $\overrightarrow{\partial}$'s on g and the parameter t has been replaced by $\frac{i\hbar}{2}$, in accordance with the physical literature. More generally, we can define a star product on $\mathbb{R}^n$ by

$$f \star g = fg + \frac{i\hbar}{2}\pi^{ij}\frac{\partial f}{\partial x^i}\frac{\partial g}{\partial x^j} + \left(\frac{i\hbar}{2}\right)^2 \pi^{ij}\pi^{kl}\frac{\partial}{\partial x_i}\frac{\partial f}{\partial x_k}\frac{\partial}{\partial x_j}\frac{\partial g}{\partial x_l} + \cdots \tag{3.27}$$

where $\{\pi^{ij}\}$ is a constant skew-symmetric tensor on $\mathbb{R}^n$ with $i, j = 1, \ldots, n$. This expression gives us the sequence of $\mathbb{R}[\![t]\!]$-bilinear bi-differential operators P_n as follows:

$$P_n(f, g) = \prod_{k=1}^{n}\pi^{i_k j_k}\left(\prod_{k=1}^{n}\frac{\partial}{\partial x_{i_k}}\right)f\left(\prod_{k=1}^{n}\frac{\partial}{\partial x_{j_k}}\right)g \tag{3.28}$$

A more elegant expression than (3.27) is given by:

$$f \star g = \langle \mathrm{d}f \otimes \mathrm{d}g, e^{\frac{i\hbar}{2}\pi}\rangle \tag{3.29}$$

or, equivalently

$$f \star g(x) = \exp\left(\frac{i\hbar}{2}\pi^{ij}\frac{\partial}{\partial x^i}\frac{\partial}{\partial y^j}\right)f(x)g(y)\Big|_{y=x}. \tag{3.30}$$

This operation defines an associative formal deformation of $\mathbb{R}^n$. Indeed we have:

$$
\begin{aligned}
((f \star g) \star h)(x) &= \exp\left(\frac{i\hbar}{2}\pi^{ij}\frac{\partial}{\partial x^i}\frac{\partial}{\partial z^j}\right)(f \star g)(x)\,h(z)\Big|_{z=x} \\
&= \exp\left(\frac{i\hbar}{2}\pi^{ij}(\frac{\partial}{\partial x^i}+\frac{\partial}{\partial y^i})\frac{\partial}{\partial z^j}\right)\exp\left(\frac{i\hbar}{2}\pi^{kl}\frac{\partial}{\partial x^k}\frac{\partial}{\partial y^l}\right) \\
&\quad \times f(x)g(y)h(z)\Big|_{x=y=z} \\
&= \exp\left(\frac{i\hbar}{2}\pi^{ij}\frac{\partial}{\partial x^i}\frac{\partial}{\partial z^j}+\pi^{kl}\frac{\partial}{\partial y^k}\frac{\partial}{\partial z^l}+\pi^{mn}\frac{\partial}{\partial x^m}\frac{\partial}{\partial y^n}\right) \\
&\quad \times f(x)g(y)h(z)\Big|_{x=y=z} \\
&= \exp\left(\frac{i\hbar}{2}\pi^{ij}\frac{\partial}{\partial x^i}(\frac{\partial}{\partial y^j}+\frac{\partial}{\partial z^j})\right)\exp\left(\frac{i\hbar}{2}\pi^{kl}\frac{\partial}{\partial y^k}\frac{\partial}{\partial z^l}\right) \\
&\quad \times f(x)g(y)h(z)\Big|_{x=y=z} \\
&= (f \star (g \star h))(x). \tag{3.31}
\end{aligned}
$$

It is evident that $P_i(f, 1) = P_i(1, f) = 0$ for any $i \geq 1$, thus the condition $f \star 1 = 1 \star f = f$ is satisfied. Putting $\{f, g\} = P_1(f, g) - P_1(g, f)$ we get the local expression of the Poisson bracket (see Eq. (2.20))

$$\{f, g\} = \pi^{ij}\frac{\partial f}{\partial x_i}\frac{\partial g}{\partial x_j}. \tag{3.32}$$

In particular, for $M = \mathbb{R}^{2n}$, using the Moyal product (3.26) we get the canonical Poisson bracket

$$\{f, g\} = \frac{\partial f}{\partial q^i}\frac{\partial g}{\partial p_i} - \frac{\partial f}{\partial p_i}\frac{\partial g}{\partial q^i}, \tag{3.33}$$

introduced in Sect. 2.1.

From the above discussion turns out that an associative (non commutative) formal deformation $\star$ of an associative commutative algebra A defines a Poisson bracket on A. We can prove that this Poisson bracket only depends on the equivalence class of $\star$, where the concept of equivalence is introduced in terms of automorphisms of $A[\![t]\!]$ as follows. Consider a generic algebra $A[\![t]\!]$ of formal power series over $k[\![t]\!]$ and two formal deformations $\star$ and $\star'$. Let J be the group of automorphisms T of $A[\![t]\!]$ such that

$$T(u) = u \qquad \mathrm{mod}\ tA[\![t]\!]. \tag{3.34}$$

for any $u \in A[\![t]\!]$.

Definition 3.3 Given two star products $\star$ and $\star'$, they are considered equivalent if there exists an element $T \in J$ such that for any $u, v \in A[\![t]\!]$

$$T(u \star v) = T(u) \star' T(v). \tag{3.35}$$

The automorphism T is determined by its restriction to A

$$T(a) = \sum_{n=0}^{\infty} T_n(a)t^n \quad a \in A, \tag{3.36}$$

where $T_i \colon A \to A$ are k-linear maps with $T_0(a) = a$. Thus, the relation (3.35) is equivalent to the set of relations

$$\sum_{i+j=n} T_i(P_j(a, b)) = \sum_{i+j+l=n} P_l'(T_i(a), T_j(b)) \quad a, b \in A. \tag{3.37}$$

In the particular case in which the algebra A is the algebra of smooth functions on M, the T_i's have to be differential operators which vanish on constants, as was proved in [30]. We denote by $[\star]$ the equivalence class of star products relative to the previous definition of equivalence. We can prove that different star products belonging to the same equivalence class induce the same Poisson bracket, simply by setting (3.16). More precisely,

Lemma 3.1 Let $\star$ be a star product on $C^\infty(M)$. The Poisson bracket

$$\{f, g\} = P_1(f, g) - P_1(g, f) \quad f, g \in C^\infty(M) \tag{3.38}$$

depends only on the equivalence class $[\star]$.

Proof Consider two equivalent star products $\star$ and $\star'$, i.e.

$$T(a \star b) = T(a) \star' T(b) \tag{3.39}$$

Expanding the formal power series of T, $\star$ and $\star'$, the term in t of this equation reads

$$P_1(f, g) + T_1(fg) = P_1'(f, g) + T_1(f)g + fT_1(g). \tag{3.40}$$

This implies that $P_1(f, g) - P_1'(f, g)$ is symmetric in f, g, hence it does not contribute to $\{f, g\}$. $\qquad \blacksquare$

This means that given an equivalence class of star products on $C^\infty(M)$, it induces a Poisson structure π on the manifold M. At this stage it is not clear whether, given a Poisson manifold, there exists a star product with the first term equal to the given Poisson structure and whether there exists a preferred choice of an equivalence class of star products. This problem has been solved by Kontsevich in [38], where he proved that there is a canonical construction of an equivalence class of star products for any Poisson manifold. More precisely, recalling the notion of equivalence classes of formal Poisson structures discussed in Sect. 2.3, we have the following

Theorem 3.1 (Kontsevich [38]). *The set of equivalence classes of star products on a smooth manifold M can be naturally identified with the set of equivalence classes of formal Poisson structures*

$$\pi = \pi_t = t\pi_1 + t^2\pi_2 + \cdots \in \mathfrak{X}^2(M)[\![t]\!], \quad [\pi, \pi]_S = 0 \qquad (3.41)$$

modulo the action of the group of formal paths in the diffeomorphisms group of M, starting at the identity diffeomorphism.

Any given Poisson structure π gives a path $\pi_t := \pi t$ and, by the above theorem, a canonical equivalence class of star products. This implies that any Poisson manifold admits a canonical formal deformation (also said canonical deformation quantization). This implication will be discussed in detail in Sect. 3.7.2.

The simplest example of a star product is the Moyal product for the Poisson structure on $\mathbb{R}^n$ with constant coefficients

$$\pi = \pi^{ij}\frac{\partial}{\partial x_i} \wedge \frac{\partial}{\partial x_j}, \quad \pi^{ij} = -\pi^{ji} \qquad (3.42)$$

for $i = 1, \ldots, n$, as discussed in Example 3.1.

3.4 Deformations and Differential Graded Lie Algebras

As mentioned above, the classification problem was solved by proving the existence of a bijection between the equivalence class of star products and the equivalence class of Poisson structures. Kontsevich proved the existence of such a bijection using the ideas of deformation theory, where deformation problems are governed by differential graded Lie algebras (DGLA). In this section we introduce the basic notion of DGLA and we discuss, roughly, how a deformation problem is attached to a DGLA. Given a DGLA L, we can define the deformations of L as the set of solutions of the Maurer–Cartan equation modulo gauge action (as it will be defined in Sect. 3.4.3).

An exhaustive review of deformation theory can be found in [40], where the deformation problem is treated by using category language. In these notes we aim to discuss the specific examples of deformation problems which are useful to prove Theorem 3.1. In particular, considering the DGLA of multivector fields, we can rephrase the equivalence class of Poisson structures in terms of Maurer–Cartan elements of this DGLA modulo gauge action (see Sect. 3.4.4). On the other hand, the equivalence class of star products can be rewritten in deformation theory by means of the DGLA of multidifferential operators.

3.4.1 Lie Algebras

The notion of Lie algebra has been already used in the previous chapter; nevertheless, as it is crucial in this section, we recall it briefly and we discuss some properties which will be useful in the discussion of differential graded Lie algebras and in particular in the definition of a gauge group. An exhaustive treatment of this subject can be found in the literature, as e.g. [12, 31].

Definition 3.4 A Lie algebra is a vector space L over a field k together with an operation $[\cdot, \cdot] : L \times L \to L$ which satisfies the following properties:

1. The bracket operation is bilinear
2. $[x, y] = -[y, x]$ (skew-symmetry)
3. $[x, [y, z]] + [y, [z, x]] + [z, [x, y]] = 0$ (Jacobi identity).

If L_1, L_2 are Lie algebras, then a linear map $\phi: L_1 \to L_2$ is a Lie algebra homomorphism if $\phi([x, y]) = [\phi(x), \phi(y)]$ for any $x, y \in L$. If ϕ is one-to-one and onto we say that ϕ is an isomorphism of Lie algebras. A Lie algebra isomorphism of L onto itself is called Lie algebra automorphism.

A linear subspace $K \subset L$ is called a Lie subalgebra if $[x, y] \in K$ for any $x, y \in K$.

Example 3.2 The space $\mathsf{End}(V)$ of all linear endomorphisms of a vector space V is a Lie algebra with bracket $[f, g] = fg - gf$. If V is finite dimensional then the subspace $sl(V) \subset \mathsf{End}(V)$ of endomorphisms with trace equal to zero is a Lie subalgebra.

For any associative algebra A we can associate a Lie algebra A_L with bracket equal to the commutator $[a, b] = ab - ba$. It is important to remark that not every Lie algebra operation is the commutator of an associative product.

Example 3.3 Let A be an associative algebra over k. The vector space of derivations of A

$$\mathsf{Der}(A, A) := \{d \in \mathsf{End}(A) : d(ab) = (da)b + adb\} \tag{3.43}$$

is a Lie subalgebra of $\mathsf{End}(A)$.

Example 3.4 Let L be a Lie algebra over k and A a commutative and k-associative algebra. Then the tensor product $L \otimes_k A$ can be made into a Lie algebra with bracket given by the bilinear extension:

$$[u \otimes a, v \otimes b] = [u, v] \otimes ab. \tag{3.44}$$

A representation of a Lie algebra L on a vector space V is a Lie algebra homomorphism $\phi: L \to \mathsf{End}(V)$. One of the most important examples of Lie algebra representations is given by the of a Lie algebra L, defined by the homomorphism

$$\mathrm{ad}(a) = [a, \cdot] : L \to \mathsf{End}(L), \mathrm{ad}(x)(y) := [x, y]. \tag{3.45}$$

A crucial notion in this context, needed to define the equivalence classes of deformations, is the exponential map. In order to introduce it and discuss its properties, we need the notion of nilpotency (the interested reader can find a more detailed discussion, with proofs and examples in [41]). First, we need the following

Definition 3.5 Let L be a Lie algebra. Using the notation

$$[U, V] = \text{span}\{[u, v] : u \in U, v \in V\}$$

the descending central series $L^{(n)}$ of L is defined as

$$L^{(1)} = L, \quad L^{(2)} = [L, L], \quad L^{(n)} = [L, L^{(n-1)}].$$

Lemma 3.2 *In the above notation we have:*

1. $L^{(n+1)} \subseteq L^{(n)}$ *for any* $n \geq 0$
2. $[L^{(i)}, L^{(j)}] \subset L^{(i+j)}$ *for any* i, j.

Definition 3.6 A Lie algebra L is called nilpotent if $L^{(n)} = 0$ for some $n \gg 0$.

If L is nilpotent, then $\text{ad}(x) \in \text{End}(L)$ is nilpotent for any $x \in L$. The converse is true when L is finite dimensional.

We remark that in general a Lie algebra is not associated to an associative product. For any nilpotent Lie algebra there exists an associative product, called Baker–Campbell–Hausdorff product, which allows us to define the group

$$\exp L = \{e^a : a \in L\} \tag{3.46}$$

of formal exponentials of elements of L. More precisely,

Theorem 3.2 *For every nilpotent Lie algebra L there is an associative product* $\bullet : L \times L \to L$ *such that*

1. *If* $f : L \to L'$ *is a morphism of nilpotent Lie algebras then*

$$f(a \bullet b) = f(a) \bullet f(b), \tag{3.47}$$

2. *If* $I \in A$ *is a nilpotent ideal of the associative algebra A and, for* $a \in I$, *we define*

$$e^a = \sum_{n=0}^{\infty} \frac{a^n}{n!} \in A; \tag{3.48}$$

then

$$e^{a \bullet b} = e^a \cdot e^b, \tag{3.49}$$

where $\cdot$ *denotes the usual product in A.*

The associative product • is defined by the formula:

$$a \bullet b = a + b + \frac{1}{2}[a, b] + \frac{1}{12}[a, [a, b]] + \frac{1}{12}[b, [a, b]] + \cdots \tag{3.50}$$

Notice that, if L is a Lie subalgebra of a nilpotent ideal of a unitary associative algebra A then (3.50) holds and the BCH product on L is also associative.

Proposition 3.1 *In the above notation*

1. for every $a, b \in A$ and $n \geq 0$

$$[a, \cdot]^n b = \sum_{i=0}^{n} (-1)^i \binom{n}{i} a^{n-i} b a^i = \sum_{i=0}^{n} \binom{n}{i} a^{n-i} b (-a)^i. \tag{3.51}$$

2. If a is nilpotent in A then also $ad(a)$ is nilpotent in $\mathsf{End}(A)$ and therefore it yields a well-defined invertible operator

$$e^{[a, \cdot]} = \sum_{n \geq 0} \frac{[a, \cdot]^n}{n!} \in \mathsf{End}(A). \tag{3.52}$$

For every nilpotent Lie algebra L there exists a natural bijection e: $L \to \exp L$ satisfying the following properties:

1. Let V be a vector space and $f : L \to \mathsf{End}(V)$ a Lie algebra homomorphism. If the image of L is contained in a nilpotent ideal, then the map

$$\exp(f) : \exp(L) \to \mathsf{Aut}(V), \quad \exp(f)(e^a) = e^{f(a)} \tag{3.53}$$

is a homomorphism of groups (here $e^{f(a)}$ denotes the usual exponential of endomorphisms).
2. If $f : L \to \mathsf{End}(V) = P$ is a representation of L as above and $[f, \cdot] : L \to \mathsf{End}(P)$ is the adjoint representation, then for every $a \in L$, $g \in \mathsf{End}(V)$

$$e^{[f, \cdot]}(e^a) g = e^{f(a)} g e^{-f(a)}. \tag{3.54}$$

3.4.2 Differential Graded Lie Algebras

In this section we aim to introduce the basic tools of differential graded Lie algebras and to give some basic examples, which are useful for the goals of this chapter. A more extensive introduction of differential graded algebras in the context of deformation theory can be found in the lecture notes by Manetti, [40, 41].

Let $\mathbb{Z}$ be the set of integers. A $\mathbb{Z}$-graded vector space, often called simply a graded vector space, is a vector space V which decomposes into a direct sum of the form

$V = \bigoplus_{n \in \mathbb{Z}} V_n$, where each V_n is a vector space. Elements of any factor V_n of the decomposition are called homogeneous elements of degree n; if $v \in V_n$, we denote $\bar{v} = \deg(v)$.

Definition 3.7 A differential graded vector space is a graded vector space $V = \bigoplus_{n \in \mathbb{Z}} V_n$ together with a linear map $d: V \to V$, called differential, such that $d(V^n) \subset V^{n+1}$ for any n and $d^2 = d \circ d = 0$.

Every complex of vector spaces (V^n, d)

$$\cdots \longrightarrow V^n \xrightarrow{d} V^{n+1} \xrightarrow{d} V^{n+2} \longrightarrow \cdots \tag{3.55}$$

can be considered as a DG vector space.

Definition 3.8 A DG (commutative) algebra is a differential graded vector space A together with a product

$$A \otimes A \to A : a \otimes b \mapsto ab, \tag{3.56}$$

which satisfies, for any $a \in A^n$, $b \in A^m$,

1. $(ab)c = a(bc)$ (associativity)
2. $ab = (-1)^{\bar{a}\bar{b}} ba$ (graded commutativity)
3. $d(ab) = d(a)b + (-1)^{\bar{a}} a d(b)$ (graded Leibniz rule)

Definition 3.9 A differential graded Lie algebra (DGLA) is a DG vector space (L, d) endowed with a bilinear operation

$$[\cdot, \cdot] : L \otimes L \to L, \tag{3.57}$$

homogeneous of degree 0, i.e. $[L^i, L^j] \subset L^{i+j}$, satisfying the following conditions, for any $a \in A^n$, $b \in A^m$:

1. $[a, b] = -(-1)^{\bar{a}\bar{b}}[b, a]$ (graded skew-symmetry)
2. $[a, [b, c]] = [[a, b], c] + (-1)^{\bar{a}\bar{b}}[b, [a, c]]$ (graded Jacobi identity)
3. $d([a, b]) = [d(a), b] + (-1)^{\bar{a}}[a, d(b)]$ (graded Leibniz rule).

Given a DGLA L, we define

$$Z^i(L) := \ker(d : L^i \to L^{i+1})$$

and

$$B^i(L) := \mathrm{Im}(d : L^{i-1} \to L^i).$$

The cohomology group associated to a DGLA L is given by

$$H^i(L) := \frac{Z^i(L)}{B^i(L)}. \tag{3.58}$$

The set $H := \bigoplus_i H^i(L)$ has a natural structure of graded Lie algebra. Indeed, it inherits the GLA structure defined on equivalence classes $|a|, |b| \in H$ by

$$[|a|, |b|]_H := |[a, b]|. \tag{3.59}$$

Finally, H is a DGLA by setting $d = 0$.

Example 3.5 Every Lie algebra is a DGLA concentrated in degree 0.

Example 3.6 Consider a DG vector space (V, d) and denote

$$\mathsf{Hom}^\bullet(V, V) = \bigoplus_{i \in \mathbb{Z}} \mathsf{Hom}^i(V, V)$$

where $\mathsf{Hom}^i(V, V) = \{f \colon V \to V \text{ linear} | f(V^n) \subset f(V^{n+i}), \ \forall n\}$. The bracket $[f, g] = fg - (-1)^{\bar{f}\bar{g}}gf$ and the differential $\delta f = [d, f] = df - (-1)^{\bar{f}} fd$ makes $\mathsf{Hom}^\bullet(V, V)$ a DGLA.

Example 3.7 Given a DGLA L and a graded commutative k-associative algebra $\mathfrak{m}$, then $L \otimes \mathfrak{m}$ has a natural structure of DGLA by setting

$$(L \otimes \mathfrak{m})^n = \oplus_i(L^i \otimes \mathfrak{m}^{n-i}),$$
$$d(x \otimes a) = dx \otimes a,$$
$$[x \otimes a, y \otimes b] = (-1)^{pr}[x, y] \otimes ab, \tag{3.60}$$

for any $a \in \mathfrak{m}^p, b \in \mathfrak{m}, x \in L$ and $y \in L^r$. Notice that, if $\mathfrak{m}$ is nilpotent, the DGLA $L \otimes \mathfrak{m}$ is also nilpotent.

Example 3.8 Let M be a differentiable manifold and $L^k := \mathfrak{X}^{k+1}(M)$. It is easy to check that the space $L := \bigoplus_{k \in \mathbb{Z}} L^k$ has a DGLA structure given by the Schouten–Nijenhuis bracket (2.30) and $d = 0$.

Example 3.9 Let A be an associative algebra over K with multiplication $m \colon A \otimes A \to A$. Denote by $L^i = \mathsf{Hom}(A^{\otimes i+1}, A)$ for any $i \geq -1$. Define

$$\circ \colon L^i \times L^j \to L^{i+j} \tag{3.61}$$

by

$$\phi \circ \psi(a_0, \dots, a_{i+j}) = \sum_s (-1)^{sj}\phi(a_0, \dots, a_{s-1}, \psi(a_s, \dots, a_{s+j}), \dots a_{i+j}). \tag{3.62}$$

It is a DGLA, called Hochschild DGLA, by setting

$$[\phi, \psi]_G = \phi \circ \psi - (-1)^{\bar{\phi}\bar{\psi}} \psi \circ \phi \tag{3.63}$$

and

$$d\phi = [m, \phi]_G.\tag{3.64}$$

The last two examples are crucial in the discussion of Kontsevich's result and will be treated more extensively in the following sections.

Definition 3.10 A morphism of DGLA is a linear homogeneous map $f: L_1 \to L_2$ of degree zero, such that

$$f \circ d = d \circ f\tag{3.65}$$

and

$$f([x, y]) = [f(x), f(y)].\tag{3.66}$$

It is important to remark that a morphism $f: L_1 \to L_2$ of DGLA's induces a morphism $H(f): H_1 \to H_2$ in cohomology; more precisely, it induces a sequence of homomorphisms $H^n(f): H^n(L_1) \to H^n(L_2)$.

Definition 3.11 A quasi-isomorphism is a morphism of DGLA's inducing isomorphisms in cohomology.

We observed that the cohomology (3.58) of a DGLA L is itself a differential graded Lie algebra with the induced bracket (3.59) and zero differential. Furthermore, we have

Definition 3.12 A differential graded Lie algebra L is formal if it is quasi-isomorphic to its cohomology, regarded as a DGLA with zero differential and the induced bracket.

This definition is crucial for stating the formality theorem; it means that the DGLA structure of a complex and a cohomology complex are preserved and the induced cohomology groups are isomorphic.

3.4.3 Maurer–Cartan Equation and Gauge Action

Eventually, we can discuss how the concept of deformation is attached to a differential graded Lie algebra via the solutions to the Maurer–Cartan equation modulo the action of a gauge group. First, we introduce the Maurer–Cartan equation of a DGLA and the gauge group and we extend them to the formal counterpart of a DGLA. As already announced, here we do not discuss the deformation theory in detail but we simply give the necessary notions to discuss concrete examples of deformation.

Definition 3.13 Given a DGLA L, the Maurer–Cartan equation of L is

$$da + \frac{1}{2}[a, a] = 0, \quad a \in L^1.\tag{3.67}$$

We denote by $MC(L) \subset L^1$ the set of solutions of the Maurer–Cartan equation. It is evident that the Maurer–Cartan equation is preserved under morphisms of differential graded Lie algebras.

One of the main goals of this section is the construction of a group which preserves the set of solutions of the Maurer–Cartan equation. This allows us to discuss the idea of how to attach deformations to DGLAs. As discussed in Sect. 3.4.1, for every Lie algebra L the set defined as $\exp L := \{e^a, a \in L\}$ can be endowed with the structure of a group via the BCH formula (3.50); it is well-defined in the case in which L is nilpotent as the infinite sum reduces to a finite one. Let us consider, in particular, the DGLA $L \otimes \mathfrak{m}$ discussed in Example 3.7 with $\mathfrak{m}$ nilpotent and under the assumption that $L^0 \otimes \mathfrak{m}$ is nilpotent (in the conventional approach $\mathfrak{m}$ is a maximal ideal in a finite-dimensional Artin ring: we do not discuss it here, but a review of functors of Artin rings can be found in [40]). Under these assumptions we can define a group G^0 simply by exponentiating the nilpotent algebra $L^0 \otimes \mathfrak{m}$. Indeed, for every $a \in L^0 \otimes \mathfrak{m}$, the corresponding adjoint operator

$$\mathrm{ad}(a) = [a, \cdot] : L \to L, \quad [a, \cdot](b) = [a, b] \tag{3.68}$$

is a nilpotent derivation of degree 0 and then its exponential

$$e^{[a,\cdot]} : L \to L, \quad e^{[a,\cdot]}b = \sum_{n=0}^{\infty} \frac{[a, \cdot]^n}{n!} b \tag{3.69}$$

is an automorphism of the DGLA $L \otimes \mathfrak{m}$. In other words,

$$G^0(L) := \exp\left(L^0 \otimes \mathfrak{m}\right) = \{\Phi : L \otimes \mathfrak{m} \to L \otimes \mathfrak{m} | \Phi = e^{[a,\cdot]}, a \in L^0 \otimes \mathfrak{m}\} \tag{3.70}$$

is a subgroup of the automorphisms of L of degree 0. As discussed above, the group structure is given by the BCH formula (3.50). In order to define the gauge action of G^0 on $L^1 \otimes \mathfrak{m}$, or more precisely on $MC(L \otimes \mathfrak{m}) \subset L^1 \times \mathfrak{m}$, we need some more properties of DGLAs.

Lemma 3.3 *If W is a linear subspace of L^i and $[L^0, L^i] \subset W$, then*

$$e^{[a,\cdot]}(v + w) = v + \underbrace{\sum_{n=1}^{\infty} \frac{1}{n!}[a, \cdot]^{n-1}([a, v]) + \sum_{n=0}^{\infty} \frac{[a, \cdot]^n}{n!} w}_{\in W} \tag{3.71}$$

for any $a \in L^0$, $v \in L^i$ and $w \in W$.

Notice also that, given a DGLA $L \otimes \mathfrak{m}$, the set $Z = \{x \in L^1 \otimes \mathfrak{m} | [x, x] = 0\}$ is stable under the adjoint action of G^0,

$$e^{[a,\cdot]}(Z) \subset Z. \tag{3.72}$$

The set Z can be related to the set of solutions of the Maurer–Cartan equation as follows. Given a DGLA $(L, [\cdot, \cdot], d)$ we can construct another DGLA L' by setting

$$(L')^i = L^i \quad \forall i \neq 1 \quad \text{and} \quad (L')^1 = L^1 \oplus kd \tag{3.73}$$

with bracket $[\cdot, \cdot]'$ defined by

$$[a + vd, b + wd]' = [a, b] + vdb + (-1)^{\bar{a}} wda \tag{3.74}$$

and differential d'

$$d'(a + vd) = [d, a + vd]' = da. \tag{3.75}$$

The natural inclusion $L \subset L'$ is a morphism of DGLA. Consider the affine embedding $\phi : L^1 \rightarrow (L')^1 : \phi(a) = a + d$; it allows us to characterize the Maurer–Cartan equation as follows

$$da + \frac{1}{2}[a, a] = 0 \Leftrightarrow [\phi(a), \phi(a)]' = 0. \tag{3.76}$$

In other words, given a DGLA L, the set of solutions of the Maurer–Cartan equations of L coincides with the set Z of L'. Furthermore, $[L^0, (L')^1] \subset L^1$ and, in particular, if L is nilpotent then also L' is nilpotent. This implies that we can apply Lemma 3.3 to L' and we can define the gauge action of $G^0(L)$ on L'_1. More precisely, in the case of $L \otimes \mathfrak{m}$ with $\mathfrak{m}$ nilpotent, we have

Definition 3.14 The gauge action of G^0 on $L^1 \otimes \mathfrak{m}$ is defined by

$$e^{[a, \cdot]} \cdot b := \phi^{-1}(e^{[a, \cdot]}\phi(b)) = e^{[a, \cdot]}(b + d) - db \tag{3.77}$$

for any $a \in L^0 \otimes \mathfrak{m}$. Explicitly, we have

$$e^{[a, \cdot]} \cdot b = \sum_{n=0}^{\infty} \frac{[a, \cdot]^n}{n!}(b) + \sum_{n=1}^{\infty} \frac{[a, \cdot]^n}{n!}(d)$$

$$= \sum_{n=0}^{\infty} \frac{[a, \cdot]^n}{n!}(b) + \sum_{n=1}^{\infty} \frac{[a, \cdot]^{n-1}}{n!}(da)$$

$$= b + \sum_{n=0}^{\infty} \frac{[a, \cdot]^n}{(n+1)!}([a, b] - da). \tag{3.78}$$

We can finally say that two elements $x, y \in L \otimes \mathfrak{m}$ are gauge equivalent if there exists $a \in L^0 \otimes \mathfrak{m}$ such that

$$y = e^{[a, \cdot]} \cdot x = x + \sum_{n=0}^{\infty} \frac{[a, \cdot]^n}{(n+1)!}([a, x] - da). \tag{3.79}$$

Since $Z = \{x \in (L')^1 \otimes \mathfrak{m} | [x, x] = 0\}$ is stable under the adjoint action of G^0 and using the characterization (3.76) of the Maurer–Cartan equation in terms of ϕ, it follows that the set of solutions of the Maurer–Cartan equation is stable under the gauge action. Now we can introduce the concept of deformation via DGLA. Usually the deformation is defined by using the notion of functor; here we try to reduce the involved notions at minimum and we say that, the deformation associated to the DGLA $L \otimes \mathfrak{m}$ is defined by

$$\mathsf{Def}_\mathfrak{m}(L) := \frac{\mathsf{MC}(L \otimes \mathfrak{m})}{G^0(L)}. \tag{3.80}$$

The meaning of this definition will become clear when we discuss two concrete examples of deformation, which will be given in the next sections. Now it is important to stress that the above discussion can be immediately extended to the formal counterpart of a DGLA, generalizing in some sense what we discussed in Sect. 2.3. Given a DGLA L over k we can define a formal counterpart $L[\![t]\!]$ over the ring $k[\![t]\!]$ of formal power series in t by

$$L[\![t]\!] := L \otimes k[\![t]\!]. \tag{3.81}$$

It has a natural structure of DGLA and the degree zero part $L^0[\![t]\!]$ is a Lie algebra. Even though it is not nilpotent we can define the gauge group formally as the set $G^0 := \exp(t L^0[\![t]\!])$, as discussed in Sect. 2.3. The action of $G^0 = \exp(t L^0[\![t]\!])$ can be defined by generalizing (3.77). More precisely,

Proposition 3.2 *Let L be a DGLA over k and let $L[\![t]\!]$ be the corresponding DGLA over $k[\![t]\!]$. Then*

$$G^0(L[\![t]\!]) = \{\Phi : L[\![t]\!] \to L[\![t]\!] | \Phi = e^{t[a,\cdot]}, \quad a \in L^0[\![t]\!]\} \tag{3.82}$$

is the subgroup of all $k[\![t]\!]$-linear automorphisms on $L[\![t]\!]$ of degree 0 which in the zeroth order of t start with Id.

Also in this case, the group structure is given by the BCH formula. In this setting we define the formal Maurer–Cartan elements as follows

Definition 3.15 Let L be a DGLA over k. An element $a \in t L^1[\![t]\!]$ is said to be a formal Maurer–Cartan element if it satisfies the Maurer–Cartan equation

$$da + \frac{1}{2}[a, a] = 0. \tag{3.83}$$

The set of formal Maurer–Cartan elements is denoted by:

$$\mathsf{MC}(L[\![t]\!]) = \{a \in t L^1[\![t]\!] | da + \frac{1}{2}[a, a] = 0\}. \tag{3.84}$$

As already stated, the action of G^0 defined in (3.82) on the set of formal Maurer–Cartan elements $MC(L[\![t]\!])$ is a direct generalization of the above discussion. Given $a \in tL[\![t]\!]$ and $g \in L^0[\![t]\!]$ the gauge action is given by

$$\exp(t\,[g, \cdot]) \cdot a := \sum_{n=0}^{\infty} \frac{(t[g, \cdot])^n}{n!}(a) - \sum_{n=0}^{\infty} \frac{(t[g, \cdot])^n}{(n+1)!}(dg)$$

$$= a + t[g, a] - t\,dg + o(t^2) \tag{3.85}$$

for any $g \in L^0[\![t]\!]$ and $a \in L^1[\![t]\!]$. Also in this case, the gauge action preserves the subset $MC(L[\![t]\!]) \subset tL^1[\![t]\!]$ of solutions of Maurer–Cartan equations.

This allows us to define the equivalence class in the formal Maurer–Cartan set as in (3.80). One can extend the deformations to algebras with linear topology which are projective limits of nilpotent algebras. Here we only remark that in the following we use $\mathfrak{m} = t\mathbb{R}[\![t]\!]$.

In the next sections we discuss two concrete examples of DGLAs and their attached deformation problems.

3.4.4 Formal Poisson Structures

The first example we aim to discuss is the DGLA of multivector fields on a smooth manifold M, already introduced in Example 3.8. We recall that, in Sect. 2.3, we introduced the concept of deformed Poisson structures and of equivalence classes of such deformations. Here we rewrite these notions by using the DGLA's approach discussed above.

Let $\mathfrak{X}^k(M)[\![t]\!]$ be the set of formal k-multivector fields on M. A formal Poisson structure π_t is a deformation of a given Poisson structure π on M which satisfies the condition

$$[\pi_t, \pi_t]_S = 0, \tag{3.86}$$

where $[\cdot, \cdot]_S$ is the Schouten–Nijenhuis bracket defined in Eq. (2.29). The equivalence class of formal Poisson structures has been introduced in Definition 2.6 by saying that two formal Poisson structures π_t and $\tilde{\pi}_t$ are equivalent if there exists a formal diffeomorphism $\phi_t = \exp \mathscr{L}_X$ with $X \in t\mathfrak{X}^1(M)[\![t]\!]$ such that

$$\pi_t = \phi_t \tilde{\pi}_t. \tag{3.87}$$

Let us consider the graded vector space (we use the notation introduced by Kontsevich)

$$T_{\text{poly}}(M) = \bigoplus_{n=-1}^{\infty} \mathfrak{X}^{n+1}(M)[\![t]\!], \tag{3.88}$$

where $\mathfrak{X}^0(M)[\![t]\!] = C^\infty(M)[\![t]\!]$ (we shift the degree by 1 in order to recover the sign used in Definition 3.9). The space $T_{\text{poly}}(M)$ can be endowed with a structure of DGLA, just considering the trivial differential d = 0 and the Schouten–Nijenhuis bracket $[\cdot, \cdot]_S$.

Now we are ready to discuss how to interpret the quotient $\text{Def}_m(L)$ defined in (3.80) for this specific DGLA. First, we observe that the set $\text{MC}(T_{\text{poly}}(M))$ of solutions of formal Maurer–Cartan equation coincides with the set of formal Poisson tensors on M. Indeed, a solution of the formal Maurer–Cartan equation is an element X of $t T_{\text{poly}}^1(M)$, a formal bi-vector field, which satisfies the following equation

$$\mathrm{d}X + \frac{1}{2}[X, X]_S = 0. \tag{3.89}$$

Since d is identically zero, the Eq. (3.89) is equivalent to

$$[X, X]_S = 0 \tag{3.90}$$

which, by Definition 2.5, is equivalent to say that X is a formal Poisson structure. Furthermore, we observe that the gauge group $G^0(T_{\text{poly}}(M))$ coincides with the group of formal diffeomorphism $\phi_t = \exp(t\mathscr{L}_X) = \exp(t[X, \cdot])$ introduced in Sect. 2.3.2. These observations allow us to claim that the quotient

$$\text{Def}(T_{\text{poly}}(M)) = \frac{\text{MC}(T_{\text{poly}}(M))}{G^0(T_{\text{poly}}(M))} \tag{3.91}$$

coincides with the equivalence class of deformations of a given Poisson structure π on M.

3.5 Deformation of Associative Algebras

In this section we aim to discuss the deformation of associative algebras via DGLA, using the notions introduced in Sects. 3.4.2 and 3.4.3. The theory of deformation of associative algebras is due to Gerstenhaber [24, 25] (see also [9, 26]). In order to give a short review of this theory we first need to introduce the DGLA structure associated to an associative algebra, the so-called Hochschild DGLA; we give an interpretation of the associated set of solutions of the Maurer–Cartan equation and its gauge equivalence and we show how the quotient (3.80) is related to the equivalence classes of formal deformations of the associative algebra that we considered. Finally, we apply this discussion to the specific case of the associative algebra of smooth functions on a manifold; in this case the equivalence class of formal deformations coincides with the equivalence class of star products defined in Sect. 3.3.

Let A be an associative algebra over a commutative ring k, with multiplication

$$\mu_0 : A \times A \to A : a \otimes b \mapsto \mu_0(a, b) = a \cdot b. \tag{3.92}$$

We recall that the algebra A is associative when the product μ_0 satisfies

$$\mu_0(\mu_0(a, b), c) = \mu_0(a, \mu_0(b, c)). \tag{3.93}$$

As discussed in Sect. 3.3, deforming an associative algebra essentially means constructing a new associative product μ which depends on a parameter t such that, in the limit $t \to 0$, μ reduces to the original product μ_0. In other words, we require that μ has the following form

$$\mu = \mu_0 + t\mu_1 + t^2\mu_2 \ldots \tag{3.94}$$

where μ_r are homomorphisms from $A \otimes A$ to A and the associativity condition (3.93) is satisfied term by term. This has been introduced in Definitions (3.1) and (3.2), where we defined a new product, called star product, by setting

$$a \star b = \mu(a, b) \tag{3.95}$$

on the algebra of formal power series $A[\![t]\!]$ over A. In order to discuss the theory of deformations of associative algebras in terms of DGLA it is useful to rewrite, using Gerstenhaber's notation, the definitions of deformation and equivalence of deformations for a generic associative algebra. This definition will generalize the notion of star product given in (3.2) and the associated concept of equivalence.

Definition 3.16 Let A be an over k with multiplication $\mu_0: A \otimes A \to A$.

1. A formal associative deformation of the multiplication μ_0 of A is a $k[\![t]\!]$-bilinear associative multiplication

$$\mu : A[\![t]\!] \times A[\![t]\!] \to A[\![t]\!] \tag{3.96}$$

    such that

$$\mu(a, b) = \mu_0(a, b), \quad \forall a, b \in A[\![t]\!] \tag{3.97}$$

    at order zero in t.

2. Two formal deformations μ and μ' of μ_0 are said to be equivalent if there exists a $k[\![t]\!]$-linear isomorphism

$$T : (A[\![t]\!], \mu) \to (A[\![t]\!], \mu'), \, T(\mu(a, b)) = \mu'(T(a), T(b)) \tag{3.98}$$

    of the form

$$T = \mathrm{Id} + \sum_{n=1}^{\infty} t^n T_n. \tag{3.99}$$

The equivalence relation (3.98) is given order by order, explicitly by the set of relations

$$\sum_{i+j=n}^{\infty} T_i(\mu_j(a, b)) = \sum_{i+j+l}^{\infty} \mu_l'(T_i(a), T_j(b)). \tag{3.100}$$

It is important to remark that a deformation of an associative algebra with unit (often said unital) is again unital, and equivalent to a deformation with the same unit (a proof can be found in [26]). Using the tools introduced in Sect. 3.4, the notions defined above can be associated to a DGLA, as the associativity can be rewritten in terms of a Maurer–Cartan equation and the equivalence in terms of a gauge group. In the following we discuss the DGLA structure that we can associate to an associative algebra and some of its basic properties, necessary for the objectives of this chapter.

3.5.1 Hochschild Complex

Let A be an associative and unital algebra. Define

$$C^{\bullet}(A) = \bigoplus_{n=-1}^{\infty} C^n(A), \text{ with } C^n(A) := \text{Hom}(A^{\otimes n+1}, A) \tag{3.101}$$

where $\text{Hom}(A^{\otimes n}, A)$ denotes the space of homomorphisms from $A \otimes \cdots \otimes A$ (n-times) to A over the ring k and we have $C^{-1}(A) = A$ (to be precise we should have used the notation $C^{\bullet}(A)[1]$, as in the standard definition $C^n(A) = \text{Hom}(A^{\otimes n}, A)$ and here we shifted each component by 1; nevertheless, in the following we do not mention the shift in order to simplify the notation). In other words, we have

$$\phi \in C^n(A), \deg \phi = n \iff \dim \phi = n + 1. \tag{3.102}$$

The differential d: $C^n(A) \to C^{n+1}(A)$ is defined by

$$(-1)^n (df)(a_0, \ldots, a_n) = a_0 f(a_1, \ldots, a_n)$$
$$- \sum_{i=0}^{n-1} (-1)^i f(a_0, \ldots, a_i a_{i+1}, \ldots, a_n) \tag{3.103}$$
$$+ (-1)^{n-1} f(a_0, \ldots, a_{n-1}) a_n$$

Definition 3.17 The Hochschild complex is the positive cochain complex $(C(A), d)$

The graded vector space $C^{\bullet}(A)$ can be endowed with a DGLA structure. First, we introduce a product operation on this complex. Consider $\phi \in C^{n-1}(A)$ and $\psi \in C^{m-1}(A)$. The Gerstenhaber product is defined by

$$\phi \circ_i \psi(a_0, \ldots, a_{n+m}) := \phi(a_0, \ldots, a_{i-1}, \psi(a_i, \ldots, a_{i+m}), a_{i+m+1}, \ldots, a_{n+m}) \tag{3.104}$$

with $i = 0, \ldots, \deg \phi$ and $a_0 \ldots a_{n+m} \in A$. Thus, we get an element $\phi \circ_i \psi \in C^{n+m}(A)$. Notice that we have

$$\deg(\phi \circ_i \psi) = \deg \phi + \deg \psi \tag{3.105}$$

for any $i = 0, \ldots, \deg \phi$. The (vector) space $C(A)$ can be endowed with the product

$$\phi \circ \psi := \sum_{i=0}^{\deg \phi} (-1)^{i \deg \psi} \phi \circ_i \psi. \tag{3.106}$$

In general this product is not associative but it allows us to introduce a Lie bracket, called Gerstenhaber bracket, defined by

$$[\phi, \psi]_G := \phi \circ \psi - (-1)^{\deg \phi \deg \psi} \psi \circ \phi. \tag{3.107}$$

Proposition 3.3 *The graded vector space $C^\bullet(A)$ with differential (3.103) and with the Gerstenhaber bracket (3.107) is a DGLA, called Hochschild DGLA.*

The proof of this Proposition can be found in [14].

Remark 3.1 It is important to remark that the Hochschild complex $(C(A)^\bullet, d)$ endowed with the Gerstenhaber bracket is not an associative algebra. For this reason, it is necessary to introduce the notion of the cup product:

Definition 3.18 Let (A, μ) an associative algebra and $\phi \in C^n(A)$, $\psi \in C^m(A)$. Then the cup product $\phi \smile \psi \in C^{n+m}(A)$ of ϕ and ψ is defined by

$$\phi \smile \psi(a_1, \ldots, a_{n+m}) = \phi(a_1, \ldots, a_n)\psi(a_{n+1}, \ldots, a_{n+m}) \tag{3.108}$$

where $a_1, \ldots, a_{n+m} \in A$.

The cup product makes $C^\bullet(A)$ into a graded associative algebra. Notice that the Hochschild differential d is a (graded) derivation of degree $+1$, i.e.

$$d(\phi \smile \psi) = d\phi \smile \phi + (-1)^n \phi \smile d\psi. \tag{3.109}$$

Finally, for $\phi \in C^n(A)$ and $\psi \in C^m(A)$ we have

$$\phi \circ d\psi - d(\phi \circ \psi) + (-1)^{m-1} d\phi \circ \psi = (-1)^{m-1}(\psi \smile \phi - (-1)^{nm} \phi \smile \psi) \tag{3.110}$$

It is important to observe that the associativity of the multiplication can be rewritten by using the Gerstenhaber bracket. More precisely, given an algebra A over k and $\mu \in C^1(A)$ a bilinear map $\mu \colon A \otimes A \to A$, it is an associative multiplication if and only if

$$[\mu, \mu]_G = 0. \tag{3.111}$$

Indeed, using the definition of Gerstenhaber bracket (3.107) we have

$$
\begin{aligned}
[\mu, \mu]_G(f, g, h) &= \sum_{i=0}^{1} (-1)^i (\mu \circ_i \mu)(f, g, h) \\
&\quad + \sum_{i=0}^{1} (-1)^i (\mu \circ_i \mu)(f, g, h) \\
&= 2(\mu(\mu(f, g), h) - \mu(f, \mu(g, h))) = 0. \tag{3.112}
\end{aligned}
$$

It is evident that condition (3.111) is equivalent to the associativity condition (3.93). Finally, we notice that the Hochschild differential can also be expressed in terms of the Gerstenhaber bracket and the multiplication $\mu : A \otimes A \to A$ of A as

$$d = [\mu, \cdot]_G : C^n(A) \to C^{n+1}(A). \tag{3.113}$$

Definition 3.19 Let (A, μ) be an associative algebra over k and $(C^\bullet(A), d, [\cdot, \cdot]_G)$ the associated Hochschild complex. The cohomology $HH^\bullet(A) = \frac{\ker d}{\operatorname{Im} d}$ is called Hochschild cohomology of (A, μ).

The interested reader is referred to [27, 33, 56].

3.5.2 Maurer–Cartan Equation and Gauge Action

In this section we discuss the deformation of an associative algebra A in terms of the associated Hochschild DGLA $(C^\bullet(A), d, [\cdot, \cdot]_G)$; we show that, also in this case, the formal deformations can be described by the set of Maurer–Cartan elements of a DGLA modulo a gauge equivalence. For this reason, we first need to reinterpret a formal deformation in terms of a Maurer–Cartan equation. More precisely, from Definition 3.16 we know that a formal deformation μ is an element of $C^1(A[[t]]) = \operatorname{Hom}(A[[t]] \otimes A[[t]], A[[t]])$, which coincides with $C^1(A)[[t]]$, that it is $k[[t]]$-bilinear, associative and that can be written in terms of formal power series as:

$$\mu = \mu_0 + \sum_{n=1}^{\infty} t^n \mu_n = \mu_0 + M \tag{3.114}$$

where M is an element of $t C^1(A)[[t]]$. As discussed above, the associativity condition, in the context of the Hochschild DGLA, is equivalent to the condition (3.111), i.e. to the set of conditions

$$[\mu_0, \mu_0]_G = 0, \tag{3.115}$$

which is automatically true as A is associative,

$$[\mu_0, \mu_1]_G + [\mu_1, \mu_0]_G = 0 \qquad (3.116)$$

and

$$\sum_{l=0}^{k} [\mu_l, \mu_{k-l}]_G = 0 \quad \forall k. \qquad (3.117)$$

In other terms, we have

$$
\begin{aligned}
[\mu, \mu]_G &= [\mu_0 + M, \mu_0 + M]_G \\
&= [\mu_0, \mu_0]_G + [\mu_0, M]_G + [M, \mu_0]_G + [M, M]_G = 0. \quad (3.118)
\end{aligned}
$$

Using the expression (3.113) for the Hochschild differential, we have $dM = [\mu_0, M]$; moreover, from the graded skew-symmetry of the Gerstenhaber bracket we have $[\mu_0, M]_G = [M, \mu_0]_G$, thus

$$[\mu, \mu]_G = 2dM + [M, M]_G. \qquad (3.119)$$

We can conclude that a deformed multiplication μ is associative if and only if $M \in tC^1(A)[\![t]\!]$ satisfies the (formal) Maurer–Cartan equation:

$$dM + \frac{1}{2}[M, M]_G = 0. \qquad (3.120)$$

Finally, we show that the equivalence classes of formal deformations can be described in terms of the gauge equivalence introduced in Proposition 3.2. More precisely, recall that two formal deformations μ and μ' are equivalent if there exists an isomorphism $T \in C^1(A)[\![t]\!]$ such that

$$T(\mu(a, b)) = \mu'(T(a), T(b)). \qquad (3.121)$$

The isomorphism is determined by

$$T = \mathrm{Id} + \sum_{n=0}^{\infty} t^n T_n. \qquad (3.122)$$

Let $T = \mathrm{Id} + \sum_{n=1}^{\infty} t^n S_n =: e^{tS} \in C^1(A)[\![t]\!]$ be the exponential function defined with respect to the Gerstenhaber product $\circ$. Then, it can be proved that, for μ and μ' in $C^1(A)[\![t]\!]$, the condition (3.121) is equivalent to

$$e^{t[S, \cdot]}(\mu) = \mu'. \qquad (3.123)$$

The details of this proof can be found in [55]. Thus, we can state the following

Proposition 3.4 *The equivalence class of formal deformations of an associative algebra A can be rewritten as the quotient $\mathsf{Def}_m(A)$ of the set of solutions of the Maurer–Cartan equation*

$$\mathrm{MC}(A) := \{M \in tC^1(A)[\![t]\!] : dM + \frac{1}{2}[M, M]_G = 0\} \tag{3.124}$$

modulo the gauge group

$$\mathrm{G}^0(A) := \{\phi : C(A)[\![t]\!] \to C(A)[\![t]\!] : \phi = e^{t[S,\cdot]}, S \in C^0(A)[\![t]\!]\}. \tag{3.125}$$

3.5.3 Star Product

In particular, we are interested to deform the algebra $C^\infty(M)$ of smooth functions on a manifold M. The formal deformations of the algebra $C^\infty(M)$ have been introduced in Sect. 3.3 by setting

$$f \star g = f \cdot g + \sum_{n=1}^{\infty} t^n P_n, \tag{3.126}$$

where $f \cdot g$ is the ordinary pointwise product of functions and the P_n's are bi-differential operators. The can be extended to elements of $C^\infty(M)[\![t]\!]$ by using the linearity over $k[\![t]\!]$ and the t-adic continuity, thus P_n's become bi-differential operators over $C^\infty(M)[\![t]\!]$. In other words, the P_n's are elements of $\mathsf{Hom}(C^\infty(M)[\![t]\!] \otimes C^\infty(M)[\![t]\!], C^\infty(M)[\![t]\!]) = C^1(C^\infty(M))[\![t]\!]$ which are bi-differential. Moreover, since we required $P_n(f, 1) = P_n(1, f) = 0$, we need to further restrict our choice by considering only differential operators which vanish on constant functions. For this reason we consider a DGL subalgebra of the Hochschild DGLA associated to $C^\infty(M)$, which we denote by $D_{\mathrm{poly}}(M)$ defined by

$$D_{\mathrm{poly}}(M) = \bigoplus_{n=1}^{\infty} D_{\mathrm{poly}}^{(n)}(M) \tag{3.127}$$

where $D_{\mathrm{poly}}^{(n)}(M) = \mathsf{Hom}(C^\infty(M)[\![t]\!]^{\otimes n+1}, C^\infty(M)[\![t]\!])$ are multidifferential operators over $C^\infty(M)[\![t]\!]$. It can be proven that $D_{\mathrm{poly}}(M)$ is a DGL subalgebra of the Hochschild DGLA as it is closed under $[\cdot, \cdot]_G$ and under the action of d. The general discussion of the previous section, which explained how we can describe formal deformations in terms of Hochschild DGLA, can be immediately applied to the subalgebra $D_{\mathrm{poly}}(M)$. Thus, we can conclude that the equivalence class of star products defined in Sect. 3.3 can be described by the set of solutions of the formal Maurer–Cartan equation $\mathrm{MC}(D_{\mathrm{poly}}(M))$ modulo the gauge equivalence.

In the next section we establish a correspondence between these DGLA's which describe the equivalence classes of formal Poisson structures and star products and we introduce the Hochschild–Kostant–Rosenberg theorem, which specifies the nature of such a correspondence. This is a fundamental step to prove Theorem 3.1, as we discuss in the last section of this chapter.

3.6 Hochschild–Kostant–Rosenberg Theorem

In the previous sections we introduced two specific DGLA's and we discussed their role in the study of deformations. The DGLA $T_{poly}(M)$ of multivector fields has been associated to the deformations of a given Poisson structure π on M and, with similar argumentation, the DGLA $D_{poly}(M)$ of multidifferential operators on M has been associated to the deformations of the associative algebra structure of smooth functions on M. In this section we introduce a result by Hochschild, Kostant and Rosenberg which proves the existence of an isomorphism between the Hochschild cohomology group of $D_{poly}(M)$ and the space of multivector fields $T_{poly}(M)$. This is crucial in the classification of quantization because it essentially means that the Poisson structure π on M is contained in a cohomology group $HH^{\bullet}(C^{\infty}(M))$ (to be precise, the cohomology group is $HH^{\bullet}_{\text{diff.v.c.}}(C^{\infty}(M))$, as we consider the sub algebra of differential operators which vanish on constant functions).

First, we observe that we can define a map

$$U_1 : T_{poly}(M) \rightarrow D_{poly}(M) \tag{3.128}$$

by setting

$$(U_1(X_0 \wedge \cdots \wedge X_n))(f_0, \ldots, f_n) = \frac{1}{(n+1)!} \sum_{\sigma \in S_{n+1}} \epsilon(\sigma) X_{\sigma(0)}(f_0) \cdots X_{\sigma(n)}(f_n)$$

$$\tag{3.129}$$

for any homogeneous element $X_0 \wedge \cdots \wedge X_n$ of $T_{poly}(M)$ (for any X_i we use the identification with differential operators) and any functions $f_0, \ldots, f_n$ on M. Here σ denotes a permutation in S_{n+1}, the set of all permutations on n elements and $\epsilon(\sigma)$ denotes the sign of this permutation. The map U_1 extends the usual identification between vector fields and differential operators and we can observe that it reduces to the identity map when applied to zero-th order vector fields.

Theorem 3.3 (Hochschild–Kostant–Rosenberg) *The map*

$$U_1 : T_{\text{poly}}(M) \to D_{\text{poly}}(M) \tag{3.130}$$

defined in (3.129) is a quasi-isomorphism of complexes.

Remark 3.2 The original result due by Hochschild, Kostant and Rosenberg has been proved in the case of smooth algebraic varieties; here we introduced the version which has been extended to the case of smooth manifolds. The reader interested in the proof of the Hochschild-Kostant-Rosenberg theorem for smooth manifolds can refer to [13]; the original proof can be found in [34].

Recalling Definition 3.11, we can claim that U_1 induces an isomorphism of the associated cohomologies $HH^\bullet(C^\infty(M))$ and $H(T_{\text{poly}}(M))$ which, in turn, coincides with $T_{\text{poly}}(M)$ itself. The Hochschild cohomology $HH^\bullet(C^\infty(M))$ carries a structure of DGLA, as remarked in Sect. 3.5.1, where the differential is identically zero and the bracket $[\cdot, \cdot]_H$ is the Gerstenhaber bracket (3.107) up to cohomology. It is important to notice that the Schouten–Nijenhuis bracket $[\cdot, \cdot]_S$ on $T_{\text{poly}}(M)$ is mapped into the bracket $[\cdot, \cdot]_H$ on the Hochschild cohomology and not into the Gerstenhaber bracket. In other words, the map U_1 is a chain map, i.e. it preserves the complex structures, but it fails to be a Lie algebra homomorphism. Indeed, we can easily check that already at order 2 we have

$$U_1([X_1 \wedge X_2, Y_1 \wedge Y_2]_S) \neq [U_1(X_1 \wedge X_2), U_1(Y_1 \wedge Y_2)]_G \tag{3.131}$$

Using the definition of Schouten–Nijenhuis bracket (2.29) we have

$$[X_1 \wedge X_2, Y_1 \wedge Y_2]_S = [X_1, Y_1] \wedge X_2 \wedge Y_2 - [X_1, Y_2] \wedge X_2 \wedge Y_1$$
$$- [X_2, Y_1] \wedge X_1 \wedge Y_2 + [X_2, Y_2] \wedge X_1 \wedge Y_1 \tag{3.132}$$

Applying this three-vector field to a triple of functions and using the map U_1 as defined in (3.129) we get

$$U_1([X_1 \wedge X_2, Y_1 \wedge Y_2]_S) = \frac{1}{6}(X_1 Y_1(f) X_2(g) Y_2(h) - Y_1 X_1(f) X_2(g) Y_2(h)$$
$$- X_1 Y_2(f) X_2(g) Y_1(h) + Y_2 X_1(f) X_2(g) Y_1(h)$$
$$- X_2 Y_1(f) X_1(g) Y_2(h) + Y_1 X_2(f) X_1(g) Y_2(h)$$
$$+ X_2 Y_2(f) X_1(g) Y_1(h) - Y_2 X_2(f) X_1(g) Y_1(h))$$
$$+ \text{perm.} \tag{3.133}$$

On the other hand, using the definition of Gerstenhaber bracket we have

$$[U_1(X_1 \wedge X_2), U_2(Y_1 \wedge Y_2)]_G = [\frac{1}{2}(X_1X_2 - X_2X_1), \frac{1}{2}(Y_1Y_2 - Y_2Y_1)]_G$$
$$\times (f \otimes g \otimes h)$$
$$= \frac{1}{4}X_1(Y_1(f)Y_2(g))X_2(h) + \cdots \qquad (3.134)$$

For this reason the map U_1 is not sufficient to build up a bijection between the equivalence classes of Poisson structures and of star products and we need to introduce a new type of morphism, whose first order approximation is the Hochschild–Kostant–Rosenberg isomorphism of complexes.

3.7 Formality Theory

Finally, we state Kontsevich's formality theorem and we discuss how this theorem is related to the quantization problem. Kontsevich's theorem extends the Hochschild–Kostant–Rosenberg map U_1 to a new kind of morphism, called L_∞-morphism, that we introduce in the following section.

3.7.1 L_∞-Morphisms of DGLA

Here we give a short review of L_∞-algebras, starting with very basic definitions, L_∞-morphisms and L_∞-quasi-isomorphisms; we introduce the L_∞-quasi-isomorphism theorem, which will be crucial for the interpretation of the formality theorem. Useful reviews on L_∞-algebras can be found in [19, 32], where they are called strong homotopy Lie algebras.

Definition 3.20 A graded coalgebra over k is a graded vector space $L = \bigoplus_{i \in \mathbb{Z}} L_i$ endowed with a coproduct, i.e. a graded linear map

$$\Delta : L \to L \otimes L \qquad (3.135)$$

such that

$$\Delta(L_i) \subset \bigoplus_{j+k=i} L_j \otimes L_k \qquad (3.136)$$

and

$$(\Delta \otimes \mathrm{Id})\Delta(x) = (\mathrm{Id} \otimes \Delta)\Delta(x) \quad \text{coassociativity} \qquad (3.137)$$

for every $x \in L$. A counit (if it exists) is a morphism

$$\varepsilon : L \to k, \qquad (3.138)$$

such that $\varepsilon(L_i = 0)$ for any $i \neq 0$ and

$$(\varepsilon \otimes \mathrm{Id})\Delta = (\mathrm{Id} \otimes \varepsilon)\Delta = \mathrm{Id}. \tag{3.139}$$

The coalgebra is cocommutative if

$$T \circ \Delta = \Delta, \tag{3.140}$$

where $T: L \otimes L \to L \otimes L$ is the twisting map

$$T(x \otimes y) := (-1)^{\overline{x}\overline{y}} y \otimes x \tag{3.141}$$

for x, y homogeneous elements of degree respectively $\bar{x}$ and $\bar{y}$.

Given a graded vector space $L = \bigoplus_{i \in \mathbb{Z}} L_i$ over k, we can define the

$$T(L) = \bigoplus_{n=0}^{\infty} L^{\otimes n} \tag{3.142}$$

with $L^{\otimes 0} = k$, and two quotients: the

$$S(L) = T(L)/\langle x \otimes y - (-1)^{\bar{x}\bar{y}} y \otimes x \rangle \tag{3.143}$$

and the

$$\Lambda(L) = T(L)/\langle x \otimes y + (-1)^{\bar{x}\bar{y}} y \otimes x \rangle; \tag{3.144}$$

these spaces are naturally graded (associative) algebras. They can be endowed with a coproduct which gives them the structure of coalgebras. In particular, on $T(L)$ the coproduct is given on homogeneous elements $v \in L$ by

$$\Delta v := v \otimes 1 + 1 \otimes v \tag{3.145}$$

and extended as an algebra homomorphism w.r.t. the tensor product. Similarly, we can define the reduced tensor algebra

$$\overline{T}(L) = \bigoplus_{n=1}^{\infty} L^{\otimes n}. \tag{3.146}$$

The projection $p: T(L) \to \overline{T}(L)$ and the inclusion $i: \overline{T}(L) \hookrightarrow T(L)$ induce a coproduct also on the reduced algebra, thus $\overline{T}(L)$ has a coalgebra structure. The reduced versions $\overline{S}(L)$ and $\overline{\Lambda}(L)$ are defined by replacing $T(L)$ by the reduced algebra $\overline{T}(L)$. Notice that also the reduced version $\overline{S}(L)$ can be endowed with the comultiplication defined above.

A differential graded coalgebra is given by a graded coalgebra endowed with the analog of a differential, called coderivation, defined in the following

Definition 3.21 A coderivation of degree n on a graded coalgebra V is a graded linear map $\delta\colon V_i \to V_{i+k}$, which satisfies the (co)-Leibniz identity

$$\Delta\delta(v) = (\delta \otimes \mathrm{Id})\Delta(v) + ((-1)^{k\bar{v}}\,\mathrm{Id} \otimes\delta)\Delta(v) \quad \forall v \in V_{\bar{v}}. \tag{3.147}$$

A differential Q on a coalgebra is a coderivation of degree one that squares to zero.

Now we can define the L_∞ structures as follows.

Definition 3.22 An L_∞-algebra is a graded vector space L over k endowed with a degree 1 coalgebra differential Q on the reduced symmetric space $\overline{S}(L[1])$.

Here we use the notation $L[1] = \bigoplus_{i\in\mathbb{Z}} L^n[1]$ with $L^n[1] := L^{n+1}$, as already mentioned in Sect. 3.5.1.

Definition 3.23 A L_∞-morphism between two L_∞-algebras, $F\colon (V, Q) \to (V', Q')$, is a morphism of graded coalgebras

$$F : C(V[1]) \to C(V'[1]) \tag{3.148}$$

such that $F \circ Q = Q' \circ F$.

It is important to remark that L_∞-algebras are in some sense a generalization of DGLA's. Indeed, we have the following

Proposition 3.5 *A L_∞-algebra is a graded vector space L endowed with a sequence of maps*

$$l_n : \wedge^n L \to L \tag{3.149}$$

of degree $2 - n$, $n > 0$, such that for every sequence of homogeneous vectors $x_1, \dots, x_n \in L$ we have

$$\sum_{k=1}^{n}(-1)^{n-k} \sum_{\sigma\in S(k,n-k)} (-1)^{\epsilon(\sigma)} l_{n-k+1}(l_k(x_{\sigma(1)}, \dots, x_{\sigma(k)}), x_{\sigma(k+1)}, \dots, x_{\sigma(n)}) = 0. \tag{3.150}$$

Here $S(k, n-k)$ is the set of unshuffles, i.e. a permutation σ such that $\sigma(i) < \sigma(i+1)$ for every $i \neq k$.

It can be checked that the sequence of maps (3.149) satisfying condition (3.150) uniquely determines a coderivation Q of degree 1. As already announced, this definition shows us that any DGLA $(L, d, [\cdot, \cdot])$ is a L_∞-algebra, by setting $l_1 = d$, $l_2 = [\cdot, \cdot]$ and $l_n = 0$ for $n > 2$.

The notion of a L_∞-morphism can also be rewritten in terms of DGLA as follows. Roughly, given two DGLA's $(L_1, d_1, [\cdot, \cdot]_1)$ and $(L_2, d_2, [\cdot, \cdot]_2)$, a L_∞-morphism $f\colon L_1 \to L_2$ is given by a sequence of linear maps

$$f_n : \wedge^n L_1^{\otimes n} \to L_2, \quad n \geq 1, \tag{3.151}$$

homogeneous of degree $1 - n$, which is compatible with the L_∞-algebra structure given by the maps (3.149). In particular, f_1 is a morphism of complexes, i.e. $f_1 \circ d_1 = d_2 \circ f_1$.

Definition 3.24 An L_∞-quasi isomorphism is an L_∞-morphism whose first component is a quasi-isomorphism.

Given a L_∞-algebra we can define a generalized Maurer–Cartan equation as

$$\sum_{n \geq 1} \frac{1}{n!} l_n(x, \ldots, x) = 0 \tag{3.152}$$

for $x \in L^1$. It is evident that, when L is a DGLA, this equation reduces to the standard Maurer–Cartan equation. The importance of this notion and of the notion of L_∞-quasi-isomorphism becomes evident from the following

Theorem 3.4 *Let* $f : L_1 \to L_2$ *be a* L_∞-*quasi isomorphism of DGLA's. Then the map*

$$x \mapsto \sum_{n \geq 1} \frac{1}{n!} l_n(x, \ldots, x) \tag{3.153}$$

induces the following bijection

$$\mathsf{Def}(L_1) \simeq \mathsf{Def}(L_2). \tag{3.154}$$

3.7.2 Formality Theorem

In this section we state the main result by Kontsevich, the so-called formality theorem, and we explain how this general result solves the classification problem introduced in Sect. 3.3.

Let us recall, briefly, the objects that are involved in this discussion. On one hand, classical mechanics is described by a Poisson structure π on a smooth manifold M, as discussed in Chap. 2. It can be always associated to a formal deformation and equivalence classes $[\pi]$ of its formal deformations coincide with the deformation $\mathsf{Def}(T_{\mathrm{poly}}(M))$ associated to the DGLA T_{poly}. On the other hand, we introduced the concept of star product in order to quantize classical systems and we could describe the equivalence class $[\star]$ of star products on $C^\infty(M)$ also in terms of the quotient $\mathsf{Def}(D_{\mathrm{poly}}(M))$. Kontsevich's formality theorem relates the complexes $D_{\mathrm{poly}}(M)$ and $T_{\mathrm{poly}}(M)$; these DGLA's are related by the map U_1 (3.129) defined in the Hochschild–Kostant–Rosenberg Theorem 3.3 but, as already pointed out, this map does not commute with the Lie bracket and for this reason it is not enough to prove the correspondence between Poisson structures and deformation quantization stated in Theorem (3.1). Kontsevich, in his famous paper [38], claimed that this defect of the U_1 map can be cured and proved that there exists a morphism between such DGLA's which preserves the Lie structure. This result is stated in the so-called formality theorem:

Theorem 3.5 (Kontsevich [38]) *There exists a natural* L_∞*-quasi isomorphism*

$$U : T_{\text{poly}}(M) \rightarrow D_{\text{poly}}(M) \tag{3.155}$$

such that the component U_1 *of* U *coincides with the quasi-isomorphism defined in the Hochschild–Kostant–Rosenberg Theorem 3.3.*

This theorem is called formality theorem, since it proves that the Hochschild DGLA $D_{\text{poly}}(M)$ and its cohomology (regarded as a DGLA) are quasi-isomorphic. In other words, it proves that the DGLA $D_{\text{poly}}(M)$ is formal (see Definition 3.12).

Kontsevich proved this theorem in [38] first considering a local version (more precisely, he studied the case for $M = \mathbb{R}^d$ and he constructed explicitly the map U) and then extending it to generic Poisson manifolds. He showed also that the L_∞-quasi isomorphism U is not unique. Notice that, unlike the map U_1, the L_∞-quasi isomorphism U preserves the DGLA structures, by definition of a L_∞-quasi isomorphism.

Finally, Kontsevich could solve the existence and classification of quantization as a corollary of his formality theorem. More precisely, he proved Theorem 3.1 that we restate here as follows,

Theorem 3.6 *Let* M *be a smooth manifold and* $C^\infty(M)$ *its algebra of smooth functions. There is a natural one-to-one correspondence between star products on* M *modulo gauge equivalence* $[\star]$ *and equivalence classes of deformations* $[\pi]$ *of the Poisson structure on* M.

This theorem is an immediate corollary of the formality theorem; indeed, we recall from Theorem 3.4 that a L_∞-quasi isomorphism of DGLA's induces a bijection of the associated equivalence classes of deformations. Thus, the L_∞-quasi isomorphism

$$U : T_{\text{poly}}(M) \rightarrow D_{\text{poly}}(M) \tag{3.156}$$

induces a bijection

$$\text{Def}(T_{\text{poly}}(M)) \simeq \text{Def}(D_{\text{poly}}(M)). \tag{3.157}$$

As recalled at the beginning of this section, the sets $\text{Def}(T_{\text{poly}}(M))$ and $\text{Def}(D_{\text{poly}}(M))$ coincide with the equivalence classes $[\pi]$ of Poisson structures and $[\star]$ of star products, respectively. This proves the one-to-one correspondence between equivalence classes of star products and equivalence classes of deformations of Poisson structures.

As discussed in Sect. 2.3.2, any Poisson structure π can be associated to a formal deformation just choosing the path $\pi_t = t\pi$. By Theorem 3.1, its equivalence class $[\pi]$ is in one-to-one correspondence with the equivalence class $[\star]$. Let us consider an element $\star$ of such an equivalence class, we have

$$\lim_{t \to 0} \frac{[f, g]_\star}{it} = \lim_{t \to 0} \frac{f \star g - g \star f}{it} = \pi(\mathrm{d}f, \mathrm{d}g) \tag{3.158}$$

Thus, the correspondence between a Poisson bracket and a commutator (with respect to the star product) is satisfied in the classical limit $t \to 0$, as announced in Sect. 3.2. Conversely, a class of star products corresponds to a class of deformations of the Poisson structure. The term in t of the star product $\star$ in such a class coincides with the term π_1 of $\pi_t = t\pi_1 + t^2\pi_2 + \cdots$. We can conclude that every classical Poisson manifold and, as a consequence every classical system, admits a canonical quantization, which is unique up to formal equivalence.

References

1. T. Ali, M. Englis, Quantization methods: a guide for physicists and analysts. Rev. Math. Phys. **17**, 391–490 (2005)
2. F. Bayen, M. Flato, C. Fronsdal, A. Lichnerowicz, D. Sternheimer, Quantum mechanics as a deformation of classical mechanics. Lett. Math. Phys. **1**, 521–530 (1977)
3. F. Bayen, M. Flato, C. Fronsdal, A. Lichnerowicz, D. Sternheimer, Deformation theory and quantization I-II. Ann. Phys. **111**(61–110), 111–151 (1978)
4. F.A. Berezin, General concept of quantization. Comm. Math. Phys. **40**, 153–174 (1975)
5. F.A. Berezin, M.A. Šubin, Symbols of operators and quantization, in *Proceedings of International Conference, Tihany, 1970*, Hilbert space operators and operator algebras, p. 1972
6. M. Bertelson, P. Bieliavsky, S. Gutt, Parametrizing equivalence classes of invariant star products. Lett. Math. Phys. **46**(4), 339–345 (1998)
7. M. Bertelson, M. Cahen, S. Gutt, Equivalence of star products. Class. Quantum Grav. **14**, A93–A107 (1997)
8. P. Bongaarts. *Quantum theory: a mathematical approach.* Springer, to appear
9. P. Bonneau, M. Flato, M. Gerstenhaber, G. Pinczon, The hidden group structure of quantum groups: strong duality, rigidity and preferred deformations. Comm. Math. Phys. **161**, 125–156 (1994)
10. M. Bordemann, Deformation quantization: a survey. ed. by J-C. Wallet. in *International Conference on Noncommutative Geometry and Physics.* vol. **103** J. Phys.: Conference Series, 2008
11. M. Bordemann, M. Brischle, C. Emmrich, S. Waldmann, Subalgebras with converging star products in deformation quantization: an algebraic construction for $\mathbb{C}P^n$. J. Math. Phys. **37**(12), 6311–6323 (1996)
12. N. Bourbaki, *Lie Groups and Lie Algebras* (Springer, Berlin, 1989)
13. M. Cahen, S. Gutt, M. de Wilde, Local cohomology of the algebra of C^∞ functions on a connected manifold. Lett. Math. Phys. **4**, 157–167 (1980)
14. A.S. Cattaneo, D. Indelicato, Poisson geometry, deformation quantization and group representation, in *Formality and Star Product*, London Mathematical Society Lecture Note series, ed. by S. Gutt, J. Rawnsley, D. Sternheimer (Cambridge University Press, Cambridge, 2004), pp. 81–144
15. L. Chloup, Star products on the algebra of polynomials on the dual of a semi-simple Lie algebra. Acad. Roy. Belg. Bull. Cl. Sci. **8**, 263–269 (1997)
16. M. De Wilde, P.B.A. Lecomte, Existence of star-products and of formal deformations of the Poisson Lie algebra of arbitrary symplectic manifolds. Lett. Math. Phys. **7**, 487–496 (1983)
17. P. Deligne, Déformations de l'Algèbre des Fonctions d'une variété symplectique: comparaison entre Fedosov et DeWilde, Lecomte. Selecta Math. N. S. **1**, 667–697 (1995)

18. P.A.M. Dirac, *Lectures on Quantum Mechanics* (Belfer Graduate School of Science, Yeshiva University, New York, 1964)
19. V. A. Dolgushev, A Proof of Tsygan's Formality Conjecture for an Arbitrary Smooth Manifold. Ph.D. thesis, MIT, 2005
20. B.V. Fedosov, A simple geometrical construction of deformation quantization. J. Differ. Geom. **40**, 213–238 (1994)
21. B.V. Fedosov, *Deformation Quantization and Index Theory* (Wiley-VCH, Weinheim, 1996)
22. M. Flato, A. Lichnerowicz, D. Sternheimer, Déformations 1-différentiales des algèbres de Lie attachées à une variété symplectique ou de contact. Compositio Math. **31**, 47–82 (1975)
23. M. Flato, A. Lichnerowicz, D. Sternheimer, Crochets de Moyal-Vey et quantification. C. R. Acad. Sci. Paris Sér. A **283**, 19–24 (1976)
24. M. Gerstenhaber, The cohomology structure of an associative ring. Ann. Math. **78**, 267–288 (1963)
25. M. Gerstenhaber, On the deformation of rings and algebras. Ann. Math. **79**(1), 59–103 (1964)
26. M. Gerstenhaber, S.D. Schack, Algebraic cohomology and deformation theory, *Deformation Theory of Algebras and Structures and Applications* (Kluwer Academic Publishers, Dordrecht, 1988), pp. 11–264
27. V.L. Ginzburg, Lectures on non commutative geometry (2005) arXiv:math/0506603
28. H.J. Groenewold, On the principles of elementary quantum mechanics. Physics **12**, 405–460 (1946)
29. S. Gutt, Deformation quantisation of Poisson manifolds. Geom. Topol. Monogr. **17**, 171–220 (2001)
30. S. Gutt, J. Rawnsley, Equivalence of star products on a symplectic manifold: an introduction to Deligne's Cech cohomology classes. J. Geom. Phys **29**, 347–392 (1999)
31. B.C. Hall, *Lie Groups, Lie Algebras and Representations* (Springer, Berlin, 2003)
32. V. Hinich, V. Schechtman, Homotopy Lie algebras, in *I.M.Gelfand Seminar*, Advances in Soviet Mathematics, ed. by S. Gindikin, vol. 16 (1993). AMS
33. G. Hochschild, On the cohomology groups of an associative algebra. Ann. Math. **46**(1), 58–67 (1945)
34. G. Hochschild, B. Kostant, A. Rosenberg, Differential forms on regular affine algebras. Trans. Amer. Math. Soc. **102**(3), 383–408 (1962)
35. A. Karabegov, Cohomological classification of deformation quantisations with separation of variables. Lett. Math. Phys. **43**, 347–357 (1998)
36. B. Keller, *Deformation quantization after Kontsevich and Tamarkin*. In: Déformation, quantification, théorie de Lie, vol. 20 (Panoramas et Synthèses, Société mathématique de France, 2005), pp. 19–62
37. M. Kontsevich, Formality conjecture, in *Deformation Theory and Symplectic Geometry*, ed. by D. Sternheimer, et al. (Kluwer Academic Publishers, Dordrecht, 1997), pp. 139–156
38. M. Kontsevich, Deformation quantization of Poisson manifolds. Lett. Math. Phys. **66**, 157–216 (2003)
39. B. Kostant, Quantization and unitary representation, *Lectures in Modern Analysis and Applications III*, Lecture Notes in Mathematics (Springer, Berlin, 1970)
40. M. Manetti. Deformation Theory Via Differential Graded Lie Algebras. Seminari di Geometria Algebrica 1998–99 Scuola Normale Superiore, 1999
41. M. Manetti, Lectures on deformations of complex manifolds. Rendiconti di Matematica **24**, 1–183 (2004)
42. A. Messiah, *Quantum Mechanics* (Dover publications, Mineola, 1999)
43. J.E. Moyal, Quantum mechanics as a statistical theory. in *Proceedings of the Cambridge Philosophical Society*, 1949, vol. 45, pp. 99–124
44. O.M. Neroslavsky, A.T. Vlasov, Sur les deformations de l'algebre des fonctions d'une variete symplectique. C. R. Acad. Sci. **292**, 71–73 (1981)
45. R. Nest, B. Tsygan, Algebraic index theorem for families. Adv. Math. **113**, 151–205 (1995)
46. H. Omori, Y. Maeda, N. Miyazaki, A. Yoshida, Deformation quantization of Fréchet-Poisson algebras of Heisenberg type. Contemp. Math. **288**, 391–395 (2001)

47. H. Omori, Y. Maeda, A. Yoshida, Weyl manifold and deformation quantization. Adv. Math. **85**(2), 224–255 (1991)

48. M. Rieffel, Deformation quantization and operator algebra. *Proceedings of Symposia in Pure Mathematics*, vol. 51, (1990)

49. M. Rieffel, Deformation quantization for actions of $\mathbb{R}^d$. Mem. Amer. Math. Soc. **506** (1993)

50. M. Rieffel, Quantization and C^*-algebras. Contemp. Math. **167**, 67–97 (1994)

51. ie segal, Quantization of nonlinear systems. J. Math. Phys. **1**(6), 339–364 (1960)

52. J-M. Souriau, *Structure des systemes dymaniques*. Editions Jacques Gabay, 1969

53. D. Sternheimer, Deformation quantization: twenty years after, in *AIP Conference Proceedings* (1998)

54. J. Vey, Déformation du crochet de Poisson sur une variété symplectique. *Commentarii Mathematici Helvetici*, **50**(1), 1975

55. S. Waldmann, *Poisson-Geometrie und Deformationsquantisierung* (Springer, Heidelberg, 2007)

56. C. Weibel, *An Introduction to Homological Algebras* (Cambridge University Press, Cambridge, 1994)

57. H. Weyl, Quantenmechanik und Gruppentheorie. Z. Physics **46**, 1–46 (1927)

58. E.P. Wigner, Quantum corrections for thermodynamic equilibrium. Phys. Rev. **40**, 749–759 (1932)

Chapter 4
Kontsevich's Formula and Globalization

In this chapter, we give a sketchy exposition of the Kontsevich formula, which allows us to define locally a star product for any Poisson manifold and we introduce some further developments of the Kontsevich theory; in particular, we introduce briefly the globalization of star products by Cattaneo-Felder-Tomassini and Dolgushev. There are many interesting problems that come up after Kontsevich's theory: in the last section of this chapter we aim to introduce some of these questions.

4.1 Kontsevich's Formula

The formality theorem was proved by Kontsevich by generalizing the explicit construction of a star product on $\mathbb{R}^d$ to any (finite-dimensional) Poisson manifold. In the following we discuss the construction of such a deformed product, the so-called Kontsevich formula, and a physical interpretation of this formula in terms of quantum field theory.

4.1.1 Kontsevich's Formula on $\mathbb{R}^d$

Let M be an open domain of $\mathbb{R}^d$, with $d \geq 1$ and let $\pi = \pi^{ij} \frac{\partial}{\partial x_i} \wedge \frac{\partial}{\partial x_i}$ be a Poisson structure in a local system of coordinates $(x_1, \ldots, x_d)$, where π^{ij} are local functions on $C^\infty(M)$. In this situation, we can write an explicit formula for the star product (modulo $O(t^3)$), which reads,

$$
\begin{aligned}
f \star g = {} & fg + t\pi^{ij} \frac{\partial f}{\partial x_i} \frac{\partial g}{\partial x_j} + \frac{t^2}{2} \pi^{ij} \pi^{kl} \frac{\partial}{\partial x_i} \frac{\partial f}{\partial x_k} \frac{\partial}{\partial x_j} \frac{\partial g}{\partial x_l} \\
& + \frac{t^2}{3} \left(\pi^{ij} \frac{\partial \pi^{kl}}{\partial x_j} \left(\frac{\partial}{\partial x_i} \frac{\partial f}{\partial x_k} \frac{\partial g}{\partial x_l} - \frac{\partial f}{\partial x_k} \frac{\partial}{\partial x_i} \frac{\partial g}{\partial x_l} \right) \right) + O(t^3). \quad (4.1)
\end{aligned}
$$

© The Author(s) 2015
C. Esposito, *Formality Theory*, SpringerBriefs in Mathematical Physics,
DOI 10.1007/978-3-319-09290-4_4

This formula reduces to the Moyal product (3.27) when the Poisson structure has constant coefficients. Here we aim to give an explicit formula for the star product for an arbitrary Poisson structure π in an open domain of $\mathbb{R}^d$. In other words, we give an explicit formula for the L_∞-quasi-isomorphism U introduced in the formality Theorem (3.5), which induces the bijection between Poisson structures and star products on $\mathbb{R}^d$. In order to write such an expression we need to find an algorithm which allows us to interpret a multivector field as a multidifferential operator. More precisely, Kontsevich formula relies on the idea of introducing a set of graphs and associating a multidifferential operator P_Γ and a weight w_Γ to each graph. The class of graphs introduced by Kontsevich is the class of oriented labeled graphs G_n, defined as follows.

Definition 4.1 An oriented graph Γ is a pair (V_Γ, E_Γ) of two finite sets such that E_Γ is a subset of $V_\Gamma \times V_\Gamma$. Elements of V_Γ are vertices of Γ and elements of E_Γ are edges of Γ. We denote by $e = (v_1, v_2)$ the edge that starts at v_1 and ends at v_2. A labeled graph Γ (sometimes called admissible graph) belongs to G_n if it satisfies the following properties:

1. Γ has $n + 2$ vertices and $2n$ edges
2. the set of vertices is decomposed in two ordered subsets, $\{1, \ldots, n\}$ and $\{L, R\}$ (where L and R are just symbols denoting left and right)
3. edges of Γ are labeled by symbols $e_1^1, e_1^2, e_2^1, e_2^2, \ldots, e_n^1, e_n^2$
4. for any $k \in \{1, \ldots, n\}$ edges labeled by e_k^1 and e_k^2 start at the vertex k
5. for any $v \in V_\Gamma$ the ordered pair (v, v) is not an edge of Γ.

For example, the graphs in Fig. 4.1 are labeled oriented graphs while, graph in Fig. 4.2 is not, because it does not satisfy condition 4. To each admissible graph $\Gamma \in G_n$ we associate a bidifferential operator

$$P_{\Gamma,\pi} : A \times A \to A, \qquad A = C^\infty(M) \tag{4.2}$$

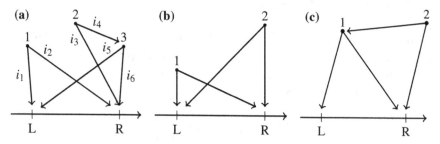

Fig. 4.1 Admissible graphs

Fig. 4.2 Non-admissible graph

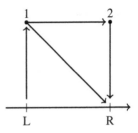

which depends on a generic bivector field $\pi \in \Gamma(\wedge^2 TM)$. The procedure to write an explicit formula for $P_{\Gamma,\pi}$ is the following:

1. we place a function f at the vertex L and a function g at R
2. we define a map $I : E_\Gamma \to \{1, \dots, d\} : (e_1^1, e_1^2, e_2^1, e_2^2, \dots, e_n^1, e_n^2) \mapsto (i_1, \dots i_d)$ so that the edges are labeled by independent indices i_l
3. for any vertex $k \in \{1, \dots, n\}$ with 2 outgoing arrows we associate the tensor $\pi^{I(e_k^1)I(e_k^2)}$
4. for any l-th arrow i_l in the set of edges ending at the vertex k we associate a partial derivative with respect to i_l acting on the function or the tensor appearing at its endpoint
5. we multiply such elements in the order prescribed by the labeling of the graph.

Following this prescription, the general formula for the operator $P_{\Gamma,\pi}$ reads,

$$P_{\Gamma,\pi} := \sum_{I:E_\Gamma \to \{1,\dots,d\}} \left[\prod_{k=1}^{n} \left(\prod_{e\in E_\Gamma, e=(\cdot,k)} \partial_{I(e)} \right) \pi^{I(e_k^1)I(e_k^2)} \right]$$
$$\times \left(\prod_{e\in E_\Gamma, e=(\cdot,L)} \partial_{I(e)} \right) f \left(\prod_{e\in E_\Gamma, e=(\cdot,R)} \partial_{I(e)} \right) g \qquad (4.3)$$

Notice that permuting the order in which we consider the edges, we get the same bidifferential operator. As an example, we compute the bidifferential operator $P_{\Gamma,\pi}$ associated to the first graph of Fig. 4.1a and we have

$$(f,g) \mapsto \sum_{i_1,\dots,i_6} \pi^{i_1 i_2} \pi^{i_3 i_4} \partial_{i_4} (\pi^{i_5 i_6}) \partial_{i_1} \partial_{i_5} (f) \partial_{i_2} \partial_{i_3} \partial_{i_6} (g). \qquad (4.4)$$

Let us define

$$U := \sum_{\Gamma \in G_n} w_\Gamma P_{\Gamma,\pi} = \sum_{\Gamma \in G_n} w_\Gamma U_\Gamma(\pi), \qquad (4.5)$$

Fig. 4.3 Angle $\phi^h(p,q)$

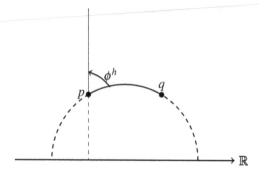

where w_Γ are certain constants in $\mathbb{R}$, called weights, that we are going to define. Kontsevich proved that there exists a choice of weights w_Γ such that $U : T_{\text{poly}}(\mathbb{R}^d) \to D_{\text{poly}}(\mathbb{R}^d) : \pi \mapsto U = \sum_{\Gamma \in G_n} w_\Gamma U_\Gamma(\pi)$ is a L_∞-quasi-isomorphism. The construction of the weights is quite hard, here we only aim to give a sketchy exposition but the interested reader can refer to [13] for a more detailed one, or [9] where the Kontsevich's formula has been derived from a path integral approach.

Let $\mathscr{H}$ be the upper half complex plane $(\text{Im}(z) > 0)$ and we endow it with the hyperbolic metric

$$ds^2 = \frac{dx^2 + dy^2}{y^2}, \tag{4.6}$$

whose geodesics are the vertical half-lines and the half-circles with center on $\mathbb{R}$. Let $p, q \in \mathscr{H}$, with $p \neq q$ and we consider the two lines $l(q, p)$ and $l(p, \infty)$, where $l(q, p)$ is the geodesic passing from p and q and $l(p, \infty)$ is the vertical line from p to infinity. The angle from $l(p, \infty)$ and $l(q, p)$ is denoted by $\phi^h(p, q)$ (h is for harmonic).

As we can see from Fig. 4.3, we have

$$\phi^h(p, q) = \arg\left(\frac{q - p}{q - \bar{p}}\right) = \frac{1}{2i} \log\left(\frac{(q - p)(\bar{q} - p)}{(q - \bar{p})(\bar{q} - \bar{p})}\right). \tag{4.7}$$

Thus, the map $(q, p) \mapsto \phi^h(p, q)$ is analytic and it admits a continuous extension to the set of pairs (q, p) such that $Im(q) \geq 0$ and $p \neq q$. Denote by $\mathscr{H}_n$ the set of n-tuples $(p_1, \ldots, p_n)$ of distinct points of $\mathscr{H}$, also called space of configurations.[1] Given a graph $\Gamma \in G_n$ and $(p_1, \ldots, p_n) \in \mathcal{H}_n$ we can represent Γ on $\mathbb{R}^2 \cong \mathbb{C}$ by associating p_i to the vertices $\{1, \ldots, n\}$ of Γ and 0 and 1 to L and R. Each arrow is represented by a geodesic segment, from its starting to its ending point, as we can see for the graph Fig. 4.1c in the following figure:

---

[1] $\mathscr{H}_n$ is a non-compact smooth $2n$-dimensional manifold and we introduce an orientation on $\mathscr{H}_n$ using its complex structure.

Fig. 4.4 Representation of Γ on $\mathbb{R}^2$

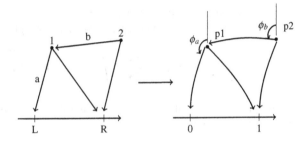

Each edge e of Γ defines an ordered pair (q, p), thus an angle $\phi_e^h := \phi^h(q, p)$ (see Fig. 4.4). We define the weight of Γ by

$$w_\Gamma := \frac{1}{n!(2\pi)^{2n}} \int_{\mathcal{H}_n} \bigwedge_{k=1}^{n} (d\phi_{e_k^1}^h \wedge d\phi_{e_k^2}^h). \qquad (4.8)$$

Lemma 4.1 [28] *The integral in the definition of w_Γ converges absolutely.*

Theorem 4.1 [28] *Let π be a Poisson bivector field in an open domain of $\mathbb{R}^d$. The formula*

$$f \star g = \sum_{n=0}^{\infty} t^n U_n(\pi)(f, g) = \sum_{n=0}^{\infty} t^n \sum_{\Gamma \in G_n} w_\Gamma P_{\Gamma,\pi}(f, g) \qquad (4.9)$$

defines an associative formal deformation (star product) of the given Poisson structure. Its equivalence class is independent of the choice of coordinates in M.

Proving that U defines an L_∞-quasi-isomorphism is far beyond this introduction; the original proof can be found in [28] and for a nice review we refer the reader to [13]. In [46, 47] Tamarkin gave a different proof of Kontsevich theorem for the case $M = \mathbb{R}^d$. Given a field k and a finite-dimensional vector field, Tamarkin proved that the shifted Hochshild complex $C(SV)[1]$ of the symmetric algebra SV, endowed with the Gerstenhaber bracket, is formal. Furthermore, it can be seen that $C(SV)[1]$ is related to $D_{\text{poly}}(M)$ by a chain of quasi-isomorphisms. Thus, Tamarkin's theorem is equivalent to the formality theorem. A review of the Tamarkin approach can also be found in [24, 25]. It is also worth to mention that Polyak [39] computed a large class of graphs, by using an interpretation of weights in terms of degree of maps; Kontsevich's star product has been studied also on the dual of a Lie algebra [45].

4.1.2 Moyal Product

In this section we aim to show that, applying formula (4.9) to the case of constant functions π^{ij}, we recover Eq. (3.27).

First, we can explicitly see that if $n = 1$ we have only two possible graphs Γ_1 and Γ_1', which differ in switching the two edges. The weight of Γ_1 is simply

$$w_{\Gamma_1} = \frac{1}{(2\pi)^2} \int_{\mathcal{H}_n} d\phi(u, 0) \wedge d\phi(u, 1). \tag{4.10}$$

Using $\phi_0 = \phi(u, 0)$ and $\phi_1 = \phi(u, 1)$ and integrating over $R = \{0 \le \phi(u, 0) \le \phi(u, 1) \le 1\}$, we get

$$w_{\Gamma_1} = \frac{1}{(2\pi)^2} \int_R d\phi_0 d\phi_1 = \frac{1}{(2\pi)^2} \frac{(2\pi)^2}{2} = \frac{1}{2}. \tag{4.11}$$

Since Γ_1 and Γ_1' differ in switching the two edges, this only implies a change of the orientation of the forms in Eq. (4.8), thus $w_{\Gamma_1} = -w_{\Gamma_1'}$. The contribution to the star product at order t is

$$\frac{t}{2} \pi^{ij} \left(\frac{\partial f}{\partial x_i} \frac{\partial g}{\partial x_j} - \frac{\partial f}{\partial x_j} \frac{\partial g}{\partial x_i} \right) = t \pi^{ij} \frac{\partial f}{\partial x_i} \frac{\partial g}{\partial x_j}, \tag{4.12}$$

as expected.

Furthermore, we observe that a graph with an edge ending in a vertex other than L or R gives zero contribution, as it includes a term of the form $\frac{\partial \pi^{jk}}{\partial x_i}$ that vanishes. At order n, we only need to consider graphs where every vertex has two edges ending in L and R. There are 2^n such graphs, differing in the order of the pair of edges starting at each vertex. As we discussed at order 1, they all have the same contribution since π^{ij} are skew-symmetric. We have a graph of order n where every vertex has the first edge to L and the second to R. Thus,

$$P_n(f, g) = 2^n w_\Gamma P_{\Gamma,\pi}(f, g) = 2^n w_\Gamma (\pi^{i_1 j_1} \ldots \pi^{i_n j_n})(\partial_{i_1} \ldots \partial_{i_n} f)(\partial_{j_1} \ldots \partial_{j_n} g). \tag{4.13}$$

In this particular case, we have $w_\Gamma = \frac{1}{n!}(w_{\Gamma_1})^n$, i.e.

$$P_n(f, g) = 2^n \frac{1}{n! 2^n} (\pi^{i_1 j_1} \ldots \pi^{i_n j_n})(\partial_{i_1} \ldots \partial_{i_n} f)(\partial_{j_1} \ldots \partial_{j_n} g). \tag{4.14}$$

We can conclude that

$$f \star g = \sum_{n=0}^{\infty} \frac{t^n}{n!} (\pi^{i_1 j_1} \ldots \pi^{i_n j_n})(\partial_{i_1} \ldots \partial_{i_n} f)(\partial_{j_1} \ldots \partial_{j_n} g), \tag{4.15}$$

which coincides with Moyal product (3.27) for $t = \frac{i\hbar}{2}$.

4.1.3 Physical Interpretation: Path Integral

The Kontsevich formula of a star product on $\mathbb{R}^d$ introduced above has been interpreted in terms of path integrals by Cattaneo and Felder in [9]. This approach involves advanced techniques of quantum field theory that we are not going to discuss here. We simply aim to explain the idea of such approach, in order to show another relation between Kontsevich's theory, purely mathematical, and physics. A complete introduction to quantum field theory can be found in classical books as [38] and [43] (the reader can also refer to the online notes by Etingof [18]). Let us recall that a field theory is a physical theory that describes the interaction of physical fields with matter. A field is a physical quantity defined at every point of the space-time; the velocity of a fluid and the electromagnetism are famous examples of classical fields. From the mathematical point of view, classical fields are described by sections of a bundle E over D, where D is a manifold representing the space-time. An observable, in this setting, can be described by a formal polynomial in the fields and their derivatives. The dynamics of the physical system is essentially encoded by the Lagrangian or, more precisely, by the integral of the Lagrangian over D; this quantity, S, is defined to be the action of the system. The principle of least action states that when a system evolves from one configuration to another, it does so along the path for which S is minimum. From this condition we get the Euler–Lagrange equations of motion for a field. The value of an observable in the system is given by its value in the solutions of Euler–Lagrange equations (also called classical solutions). It is important to remark that classical fields can not describe quantum mechanical aspects of physical phenomena: for instance, it is known that electromagnetism has also a quantum nature, since certain aspects of the behavior of the light involve discrete particles rather than fields. This clarifies the necessity of a quantum field theory; the quantization of a field theory can be performed by two different approaches, called canonical quantization and path integral formulation. The path integral has been introduced in the study of a quantum particle motion to evaluate the correlation functions, which are given by integrals that can not be handled rigorously. For this reason, they have been defined in perturbation theory, as formal series in $\hbar$. In other words, the path integral formulation replace the notion of a single trajectory with an integral over an infinite number of possible trajectories. This formulation is very useful for the development of quantum field theory; the expected value of an observable is given, in this setting, by the integral over all possible classical field configurations with a phase given by the classical action evaluated in that field configuration.

The Kontsevich formula has been interpreted in terms of a particular field theory: the authors in [9] describe the quantization of such a field theory explicitly and they show that Kontsevich's formula is given by the perturbative expansion of the path integral. Roughly, they consider two bosonic fields on a disc D, X and η. X is a map from D to a Poisson manifold M and η is a differential one-form on D. The star product of two functions f and g on M, at $x \in M$, is given by the semiclassical expansion of the path integral:

$$f \star g(x) = \int_{X(\infty)=x} f(X(1))g(X(0))e^{\frac{i}{\hbar}S[X,\eta]}dXd\eta, \qquad (4.16)$$

where 0, 1, ∞ are three points on the unit circle and $S[X, \eta]$ is the action. The path integral is over all X and η and its semiclassical expansion is an expansion around the classical solution $X(u) = x$, $\eta(u) = 0$ for $u \in D$. The evaluation of such path integral is quite complicated, due to the presence of a particular gauge symmetry. The gauge transformation, in this case, forms a Lie algebra only when acting on classical solutions and this makes the evaluation of the path integral much harder, since the usual method (the so-called BRST method) does not work. The generalization that works in this case is the Batalin-Vilkovisky method (see [1, 2] for details on the theory), which yields to a gauge fixed action. Using such an action, Cattaneo and Felder computed the Feynman perturbation expansion in power of $\hbar$ around the classical solutions and they could show that it reproduces Kontsevich's formula.

4.2 Globalization

The formality theorem, discussed in the previous chapter, has been first proved by Kontsevich in [28] by extending the quantization obtained in the case $M = \mathbb{R}^d$ to a general Poisson manifold. Kontsevich's proof is very complicated and we do not discuss it here. The globalization of the quantization has been proved with different approaches by Cattaneo et al. in [10, 12] and, more recently, by Dolgushev [16, 17]. In the following we briefly introduce both theories, minimizing the technical aspects.

4.2.1 The Approach of Cattaneo–Felder–Tomassini

In this section we give a short review of the works of Cattaneo et al. [10] (see also [11–13]), where the authors give an explicit construction of a star product on any Poisson manifold. This construction is similar, in the spirit, to Fedosov deformation quantization of symplectic manifolds [19]. For this reason, it is useful to recall the guidelines of Fedosov construction; the main idea of Fedosov was to construct a star product on a symplectic manifold by identifying the space $C^\infty(M) [\![t]\!]$ with the algebra of flat sections of the so-called Weyl bundle endowed with a flat connection. The first step consists in the construction of a vector bundle W associated to a symplectic manifold. More precisely, given a symplectic manifold (M, ω), ω defines a symplectic structure on each tangent space $T_x M$ and this allows us to construct a corresponding associative algebra W_x (called formal Weyl algebra), where the elements are formal power series and the product is given by the Weyl rule (3.30). Taking the union of such algebras W_x, $x \in M$, we obtain a bundle W of formal Weyl algebras. As a second step, Fedosov defined a general connection D on W and its curvature Ω and he proved some important properties for D and Ω (they are generally called Fedosov connection and Weyl curvature, resp.). Finally, Fedosov defined a new connection $\overline{D}$ obtained by deforming D and he proved that $\overline{D}$ satisfies the same properties of D and it is flat. This implies that there is a bijection between $C^\infty(M) [\![t]\!]$ and the space $W_{\overline{D}}$ of flat sections on W w.r.t. $\overline{D}$ and the Weyl product on $W_{\overline{D}}$ can be transported to $C^\infty(M) [\![t]\!]$ yielding a star product.

The globalization introduced by Cattaneo, Felder and Tomassini in the Poisson case is quite similar but the techniques involved to construct a vector bundle associated to a Poisson manifold are quite hard; here we only aim to give an outline, addressing the reader to the original paper [10] for a complete exposition. The main idea in this paper is to realize the deformed algebra $C^\infty(M) [\![t]\!]$ as the algebra of horizontal sections of a bundle of algebras. Thus, the first step is the construction of such a bundle on the Poisson manifold (M, π). First, we define the *classical* bundle $E_0 \to M$ which is a Poisson algebra bundle, i.e. a vector bundle whose fibers are Poisson algebras. The vector bundle E_0 can be endowed with a canonical flat connection D_0 and we denote by $H^0(E_0, D_0)$ the space of D_0-horizontal sections of E_0 (sections which are constant along smooth paths in M). Notice that the canonical map $C^\infty(M) \to E_0$ is a Poisson algebra isomorphism onto $H^0(E_0, D_0)$, thus $C^\infty(M) \cong H^0(E_0, D_0)$. The second bundle is somehow the quantum version of E_0. Its construction is very technical and it needs some basis of formal geometry, that we are not going to discuss here. We just remark that formal geometry studies infinite-dimensional manifolds of jet spaces and it is useful for the globalization of the Kontsevich's formula as it allows us to describe the global behavior of objects defined locally in terms of coordinates. We introduce the second bundle E roughly, as a bundle of associative algebras over $\mathbb{R} [\![t]\!]$, which can be obtained by quantizing the fibers of E_0. Its construction depends on the choice of the coordinate system ϕ_x, $x \in M$. The fibers of E are endowed with an associative product, defined by applying Kontsevich's formula for $\mathbb{R}^d$ w.r.t. the coordinate system ϕ_x. The details of this construction are discussed in [10], where the authors also give a short review of formal geometry as it was introduced by Gelfand and Kazhdan in [20]. A nice introduction to the language of jets can be found in [27] (also useful to study basic differential geometry) and a review of bundles of infinite jets can be found in [44]. Furthermore, formal geometry has been exhaustively treated in [8].

As in the Fedosov construction, now we need to define a connection on E and we can see that the Kontsevich's formula for $\mathbb{R}^d$ provides the ingredients to construct such a connection. Recall that, given a Poisson structure π, the Kontsevich star product is given by

$$f \star g = fg + \sum_{k=1} t^k U_k(\pi)(f, g). \tag{4.17}$$

More generally, considering $U_n(\pi_1, \ldots, \pi_j)$ defined as a multilinear graded symmetric function of j multivector fields with values in multidifferential operators on $C^\infty(\mathbb{R}^d)$, the formality theorem can be rewritten as follows:

Theorem 4.2 *Let $\pi_j \in \Gamma(\wedge^{m_j} T\mathbb{R}^d)$, $j = 1, \ldots, n$ be multivector fields. Let $\varepsilon_{ij} = (-1)^{(m_1 + \ldots m_{i-1})m_i + (m_1 + \ldots m_{i-1} + m_{i+1} + \ldots + m_{j-1})m_j}$. Then, for any functions $f_0, \ldots, f_m$,*

$$\sum_{l=0}^{n} \sum_{k=-1}^{m} \sum_{i=0}^{m-k} (-1)^{k(i+1)+m} \sum_{\sigma \in S_{l,n-l}} \varepsilon(\sigma) U_l(\pi_{\sigma(1)}, \ldots, \pi_{\sigma(l)})(f_0 \otimes \ldots \otimes f_{i-1}$$

$$\otimes U_{n-l}(\pi_{\sigma(l+1)}, \ldots, \pi_{\sigma(n)})(f_i \otimes \ldots \otimes f_{i+k}) \otimes f_{i+k+1} \otimes \ldots \otimes f_m)$$

$$= \sum_{i<j} \varepsilon_{ij} U_{n-1}([\pi_i, \pi_j]_S, \pi_1, \ldots, \hat{\pi}_i, \ldots, \hat{\pi}_j, \ldots, \pi_n)(f_0 \otimes f_m). \quad (4.18)$$

Here we denoted by $S_{l,n-l}$ the subset of the group S_n of permutations of n letters consisting of permutations such that $\sigma(1) < \ldots < \sigma(l)$ and $\sigma(l+1) < \ldots < \sigma(n)$. For $\sigma \in S_{l,n-l}$ let

$$\varepsilon(\sigma) = (-1)^{\sum_{r=1}^{l} m_{\sigma(r)} (\sum_{s=1}^{\sigma(r)-1} m_s - \sum_{s=1}^{r-1} m_{\sigma(s)})}. \quad (4.19)$$

Consider some special case of this theorem, namely the cases involving a Poisson bi-vector field π and two vector fields X and Y. Let us introduce:

$$P(\pi) = \sum_{k=0}^{\infty} \frac{t^k}{k!} U_k(\pi),$$

$$A(X, \pi) = \sum_{k=0}^{\infty} \frac{t^k}{k!} U_{k+1}(X, \pi),$$

$$F(X, Y, \pi) = \sum_{k=0}^{\infty} \frac{t^k}{k!} U_{k+2}(X, Y, \pi). \quad (4.20)$$

It is evident that, the coefficients of P are bi-differential operators (as in the usual formulation of formality for $\mathbb{R}^d$), while the coefficients of A and F are differential operators and functions, respectively. They satisfy the relations of Theorem 4.2. In other words, P, A, and F are elements of degree 0, 1 and 2 resp., of the Lie algebra cohomology complex[2] of (formal) vector fields with values in the space of multi-differential operators depending polynomially on π (the so-called local polynomial maps). We denote by $\mathfrak{U}$ the space of these local polynomial maps and since the Lie algebra W of vector fields on $\mathbb{R}^d$ acts on $\mathfrak{U}$ we can form a Lie algebra cohomology complex $C^\bullet(W, \mathfrak{U}) = \operatorname{Hom}_{\mathbb{R}}(\wedge^\bullet W, \mathfrak{U})$. An element S of $C^k(W, \mathfrak{U})$ is a map which sends $X_1 \wedge \ldots \wedge X_k$ to a multidifferential operator $S(X_1, \ldots, X_k, \pi)$.

---

[2] The definition of Lie algebra cohomology is recalled in Appendix A.

The differential on this complex is defined by

$$
\begin{aligned}
\delta S(X_1, \dots, X_k + 1, \pi) := & \sum_{i=1}^{k+1} (-1)^i \left. \frac{d}{dt} \right|_{t=0} S(X_1, \dots, \hat{X}_i, \dots, X_{k+1}, (\Phi_X^t)_* \pi) \\
& + \sum_{i<j} S([X_i, X_j], X_1, \dots, \hat{X}_i, \dots, \hat{X}_j, x, \dots, X_{k+1}, \pi),
\end{aligned}
$$

$$(4.21)$$

where Φ_X^t denotes the flow of the vector field X. From the formality theorem, using the definitions (4.20), we get relations for P, A and F; in particular, the relations obtained for P are the defining conditions of a star product and the ones involving A are used to construct a connection D on E. It can be seen that the space $\Gamma(E)$ of sections of E can be endowed with a deformed product on $\star$ which, as the Weyl product in the Fedosov construction, will give us the deformed product on $C^\infty(M)$. More precisely, we identify E with the trivial bundle with fiber $\mathbb{R}[[y^1, \dots, y^d, t]]$ (this is crucial; in this way E realizes the desired quantization since we can assume it is isomorphic to the bundle $E_0[[t]]$, whose elements are formal power series with infinite jets as coefficients). A section $f \in \Gamma(E)$ is given by a local map $x \mapsto f_x$ where for any y, $f_x(y)$ is a formal power series in $\mathbb{R}[[y^1, \dots, y^d, t]]$ and the product of two sections f and g is given by $(f \star g)_x = P(\pi_x)(f_x, g_x)$; here π_x is the pushforward by ϕ_x^{-1} of the Poisson structure π on $\mathbb{R}^d$ and we get

$$
(f \star g)_x(y) = f_x(y)g_x(y) + t \sum_{i,j=1}^{d} \pi_x^{ij}(y) \frac{\partial f_x}{\partial y^i}(y) \frac{\partial g_x}{\partial y^j}(y) + \dots \tag{4.22}
$$

Now we can define the connection D on $\Gamma(E)$ by setting

$$
(Df)_x = d_x f + A_x^M f, \tag{4.23}
$$

where $d_x f$ is the de Rham differential of f, viewed as a function of $x \in M$ with values in $\mathbb{R}[[y^1, \dots, y^d, t]]$ and for $X \in T_x M$

$$
A_x^M(X) = A(\hat{X}_x, \pi_x) \tag{4.24}
$$

where A is the operator defined in Eq. (4.20) evaluated on the multivector fields X and π expressed in the local coordinate system ϕ_x.

It has been proven in [10] that D induces a global connection on E because, from the properties of A, we can see that D is independent of the choice of the local coordinates. The connection D is defined on the space of formal one differential forms $\Omega^1(E) = \Omega^1(M) \otimes_{C^\infty(M)} \Gamma(E)$ and can be extended to the whole complex $\Omega^\bullet$; using the properties of A and F we have the following

Lemma 4.2 [10] *Let* $F^M \in \Omega^2(E)$ *be the E-valued two form* $x \mapsto F_x^M$, *with* $F_x^M(X, Y) = F(\hat{X}_x, \hat{Y}_x, \pi_x)$, *with* $X, Y \in T_x M$. *Then for any f and g in $\Gamma(E)$*

1. $D(f \star g) = Df \star g + f \star Dg$
2. $D^2 f = F^M \star f - f \star F^M = [F^M, f]_\star$
3. $DF^M = 0$

A connection D satisfying these identities on a bundle E of associative algebras is called Fedosov connection and its curvature F is called Weyl curvature. Notice that, from identity 1 of Lemma 4.2, the space of horizontal sections $\ker D$ forms an algebra but D is not flat. Thus, also in this setting, we need to deform D in such a way we get a flat connection still preserving identity 1 of Lemma 4.2. This allows us to deform the algebra of horizontal sections of E with respect to the deformed connection.

First, we can see that given a Fedosov connection D we can deform it and get another Fedosov connection $\overline{D}$; more precisely,

Proposition 4.1 [10] *If D is a Fedosov connection and $\gamma \in \Omega^1(E)$, then $\overline{D} = D + [\gamma, \cdot]_\star$ is a Fedosov connection with curvature $\overline{F} = F + D\gamma + \gamma \star \gamma$.*

Recall that a Fedosov connection is flat if $D^2 = 0$; in this case, we can define the cohomology groups

$$H^i(E, D) = \frac{\ker(D : \Omega^i(E) \to \Omega^{i+1}(E))}{Im(D : \Omega^{i-1}(E) \to \Omega^i(E))}. \tag{4.25}$$

Given the vector bundle E_0, let $E_0[\![t]\!]$ be the formal counterpart. Sections of $E_0[\![t]\!]$ are formal power series in t with coefficients in $\Gamma(E_0)$. Assume that $E = E_0[\![t]\!]$ and that D is a Fedosov connection on E with Weyl curvature F. They can be expanded as formal power series

$$D = D_0 + tD_1 + \dots \tag{4.26}$$

and

$$F = F_0 + tF_1 + \dots \tag{4.27}$$

where D_0 is a Fedosov connection on the bundle of algebras E_0 with Weyl curvature F_0.

Lemma 4.3 *If $F_0 = 0$ and $H^2(E_0, D_0) = 0$ there exists a $\gamma \in t\Omega^1(E)$ such that $\overline{D} = D + [\gamma, \cdot]_\star$ is flat. As we consider the classical bundle E_0 with canonical flat connection D_0, this implies that the deformed connection $\overline{D}$ (which is Fedosov by Proposition 4.1) is flat.*

Thus, $H^0(E, \overline{D}) = \ker \overline{D}$ is an algebra over $\mathbb{R}[\![t]\!]$ and there is an isomorphism $\rho : H^0(E, \overline{D}) \to H^0(E_0, D_0)$. Since $H^0(E_0, D_0) \cong C^\infty(M)$ we can map the star product (4.22) on $C^\infty(M)$ by means of the isomorphism ρ; the authors in [10] proved that the product obtained is a well defined star product which deforms the Poisson structure π on M.

4.2.2 Dolgushev's Construction

Another interesting, and more general, approach to prove the formality theorem in its global version is due to Dolgushev [16, 17]; his approach is more general as he proved the formality theorem for a generic manifold (Kontsevich in [28] already gave a sketchy proof of formality for arbitrary manifolds). The basic idea is to construct Fedosov resolutions of the algebras of multivector fields and multidifferential operators which allow us to extend formality theorem for $\mathbb{R}^d$ fiberwise.

The first step of this construction consists in defining a new bundle $\mathcal{S}M$ which is a natural analogue of the Weyl algebra bundle used by Fedosov. The bundle $\mathcal{S}M$ is defined as the (formally completed) symmetric algebra of the cotangent bundle T^*M; more precisely,

Definition 4.2 The bundle $\mathcal{S}M$ is a bundle over the manifold M, whose sections are in the form

$$a = a(x, y) = \sum_{p=0}^{\infty} a_{i_1 \dots i_p}(x) y^{i_1} \dots y^{i_p}, \tag{4.28}$$

where $a_{i_1 \dots i_p}(x)$ are symmetric covariant tensors in the local coordinates x^i and y^i are variables which transform as contravariant vectors (for this reason they can be interpreted as formal coordinates on the fibers of TM). The indices $i_1, \dots, i_p$ run from 1 to d.

The space $\Gamma(\mathcal{S}M)$ of sections on $\mathcal{S}M$ is a commutative algebra with a unit, as we can endow it with the product induced by a fiberwise multiplication of formal power series in y^i. Now we can introduce (formal fiberwise) multivector fields and multidifferential operators on $\mathcal{S}M$.

Definition 4.3 A bundle $\mathcal{T}_{\text{poly}}^k$ of formal fiberwise multivector fields of degree k is a bundle over M whose sections $v : \wedge^{k+1} \Gamma(\mathcal{S}M) \to \Gamma(\mathcal{S}M)$ are linear operators on $C^\infty(M)$ of the form

$$v = \sum_{p=0}^{\infty} v_{i_1 \dots i_p}^{j_0 \dots j_k}(x) y^{i_1} \dots y^{i_p} \frac{\partial}{\partial y^{j_0}} \wedge \dots \wedge \frac{\partial}{\partial y^{j_k}}, \tag{4.29}$$

where the infinite sum in y's is formal and $v_{i_1 \dots i_p}^{j_0 \dots j_k}(x)$ are tensors symmetric in $i_1, \dots, i_p$ and antisymmetric in $j_0, \dots, j_k$.

Then, the total bundle $\mathcal{T}_{\text{poly}}$ is given by

$$\mathcal{T}_{\text{poly}} = \bigoplus_{k=-1}^{\infty} \mathcal{T}_{\text{poly}}^k, \qquad \mathcal{T}_{\text{poly}}^{-1} = \mathcal{S}M \tag{4.30}$$

Similarly, we define the bundle of multidifferential operators.

Definition 4.4 A bundle $\mathcal{D}^k_{\text{poly}}$ of formal fiberwise multidifferential operators of degree k is a bundle over M whose sections are $C^\infty(M)$-multilinear maps B: $\otimes^{k+1}\Gamma(\mathcal{S}M) \to \Gamma(\mathcal{S}M)$ of the form

$$B = \sum_{\alpha_0 \ldots \alpha_k} \sum_{p=0}^{\infty} B^{\alpha_0 \ldots \alpha_k}_{i_1 \ldots i_p}(x) y^{i_1} \ldots y^{i_p} \frac{\partial}{\partial y^{\alpha_0}} \otimes \ldots \otimes \frac{\partial}{\partial y^{\alpha_k}}, \tag{4.31}$$

where α's are multi-indices $\alpha = j_1 \ldots j_l$ and

$$\frac{\partial}{\partial y^\alpha} = \frac{\partial}{\partial y^{j_1}} \ldots \frac{\partial}{\partial y^{j_l}}; \tag{4.32}$$

the infinite sum in y's is formal, and the sum in the orders of derivatives $\frac{\partial}{\partial y}$ is finite.

The total bundle $\mathcal{D}_{\text{poly}}$ is given by

$$\mathcal{D}_{\text{poly}} = \bigoplus_{k=-1}^{\infty} \mathcal{D}^k_{\text{poly}}, \qquad \mathcal{D}^{-1}_{\text{poly}} = \mathcal{S}M \tag{4.33}$$

Finally, we need to consider the graded-commutative algebra $\Omega(M, \mathcal{S}M)$ of exterior forms on M with values in $\mathcal{S}M$,

$$\Omega(M, \mathcal{S}M) = \{a(x, y, dx) = \sum_{p, q \geq 0} a_{i_1 \ldots i_p j_1 \ldots j_q}(x) y^{i_1} \ldots y^{i_p} dx^{j_1} \ldots dx^{j_q}\},$$
$$\tag{4.34}$$

where $a_{i_1 \ldots i_p j_1 \ldots j_q}(x)$ are contravariant tensors symmetric in the indices $i_1, \ldots, i_p$ and anti-symmetric in $j_1, \ldots, j_q$. Similarly, we can define the vector spaces $\Omega(M, \mathcal{T}_{poly})$ and $\Omega(M, \mathcal{D}_{poly})$ of smooth exterior forms on M with values in $\mathcal{T}_{poly}$ and $\mathcal{D}_{poly}$ respectively. They are DGLA's and we denote by d and $[\cdot, \cdot]_G$ the differential and the Lie bracket in $\Omega(M, \mathcal{D}_{poly})$ and by $[\cdot, \cdot]_S$ the Lie bracket on $\Omega(M, \mathcal{T}_{poly})$ (recall that the differential on $\mathcal{T}_{poly}$ is identically zero). Notice that the fibers of $\mathcal{T}_{poly}$ and $\mathcal{D}_{poly}$ form a DGLA, $\mathcal{T}_{poly}(\mathbb{R}^d)$ and $\mathcal{D}_{poly}(\mathbb{R}^d)$ respectively (more precisely, on the formal completion of $\mathbb{R}^d$), so the formality for $\mathbb{R}^d$ implies that we have a fiberwise L_∞ quasi-isomorphism

$$U^f : (\Omega(M, \mathcal{T}_{poly}), 0, [\cdot, \cdot]_S) \to (\Omega(M, \mathcal{D}_{poly}), d, [\cdot, \cdot]_G). \tag{4.35}$$

Using the same technique as Fedosov, Dolgushev deformed the above DGLA's and proved that there exists L_∞-quasi-isomorphism from $\mathcal{T}_{poly}$ and the deformed complex associated to $\Omega(M, \mathcal{T}_{poly})$ (similarly, from $\mathcal{D}_{poly}$ to the deformed complex associated to $\Omega(M, \mathcal{D}_{poly})$). This allows us to prove that there exists a L_∞-quasi-isomorphism from $\mathcal{T}_{poly}$ to $\mathcal{D}_{poly}$ for any smooth manifold M. Let us define a differential operator on the algebra $\Omega(M, \mathcal{S}M)$ by

$$\delta = dx^i \frac{\partial}{\partial y^i} : \Omega^q(M, \mathcal{S}M) \to \Omega^{q+1}(M, \mathcal{S}M), \qquad \delta^2 = 0. \tag{4.36}$$

This differential extends to differentials on $\Omega(M, \mathcal{T}_{poly})$ and $\Omega(M, \mathcal{D}_{poly})$ as follows

$$\delta = \left[dx^i \frac{\partial}{\partial y^i}, \cdot \right]_S : \Omega^q(M, \mathcal{T}_{poly}) \to \Omega^{q+1}(M, \mathcal{T}_{poly}), \qquad \delta^2 = 0 \tag{4.37}$$

and

$$\delta = \left[dx^i \frac{\partial}{\partial y^i}, \cdot \right]_G : \Omega^q(M, \mathcal{D}_{poly}) \to \Omega^{q+1}(M, \mathcal{D}_{poly}), \qquad \delta^2 = 0. \tag{4.38}$$

It is easy to check that δ is compatible with the DGLA structures on $\Omega(M, \mathcal{T}_{poly})$ and $\Omega(M, \mathcal{D}_{poly})$.

The zero cohomologies of the complexes $(\Omega(M, \mathcal{S}M), \delta)$, $(\Omega(M, \mathcal{T}_{poly}), \delta)$ and $(\Omega(M, \mathcal{D}_{poly}), \delta)$ can be computed easily and it turns out that

$$H^0(\Omega(M, \mathcal{S}M), \delta) = C^\infty(M) \tag{4.39}$$

and

$$H^0(\Omega(M, \mathcal{T}_{poly}), \delta) = \mathcal{F}^0\mathcal{T}_{poly}, \qquad H^0(\Omega(M, \mathcal{D}_{poly}), \delta) = \mathcal{F}^0\mathcal{D}_{poly}, \tag{4.40}$$

where $\mathcal{F}^0\mathcal{T}_{poly}$ is just the vector space of all fiberwise multivector fields (4.29) and $\mathcal{F}^0\mathcal{D}_{poly}$ is the vector space of all fiberwise multidifferential operators (4.31). Now we need to deform the above complexes in such a way we can identify $\mathcal{F}^0\mathcal{T}_{poly}$ with $\mathcal{T}_{poly}$ and $\mathcal{F}^0\mathcal{D}_{poly}$ with $\mathcal{D}_{poly}$. Let us introduce an affine torsion free connection ∇_i on M and associate to it the following derivation of $\Omega(M, \mathcal{S}M)$

$$\nabla = dx^i \frac{\partial}{\partial x^i} + \Gamma : \Omega^q(M, \mathcal{S}M) \to \Omega^{q+1}(M, \mathcal{S}M), \tag{4.41}$$

where

$$\Gamma = -dx^i \Gamma_{ij}^k(x) y^j \frac{\partial}{\partial y^k}, \tag{4.42}$$

with $\Gamma_{ij}^k(x)$ being the Christoffel symbols of ∇_i. The derivation ∇ extends to derivations of the DGLA's $\Omega(M, \mathcal{T}_{poly})$ and $\Omega(M, \mathcal{D}_{poly})$ as follows

$$\nabla = dx_i \frac{\partial}{\partial x^i} + [\Gamma, \cdot]_S : \Omega^q(M, \mathcal{T}_{poly}) \to \Omega^{q+1}(M, \mathcal{T}_{poly}) \tag{4.43}$$

and

$$\nabla = dx_i \frac{\partial}{\partial x^i} + [\Gamma, \cdot]_G : \Omega^q(M, \mathcal{D}_{poly}) \to \Omega^{q+1}(M, \mathcal{D}_{poly}). \tag{4.44}$$

In general ∇ is not flat but Dolgushev proved that it can be used to deform the differential δ in such a way we get a flat derivation D. In particular,

$$D = \nabla - \delta + A : \Omega^q(M, \mathcal{S}M) \to \Omega^{q+1}(M, \mathcal{S}M),$$
$$D = \nabla - \delta + [A, \cdot]_S : \Omega^q(M, T_{\text{poly}}) \to \Omega^{q+1}(M, T_{\text{poly}}),$$
$$D = \nabla - \delta + [A, \cdot]_G : \Omega^q(M, \mathcal{D}_{\text{poly}}) \to \Omega^{q+1}(M, \mathcal{D}_{\text{poly}}). \quad (4.45)$$

It has been proven that there exists a suitable A, which makes the derivation D flat; for this reason D is called Fedosov differential. The new complexes $(\Omega(M, \mathcal{S}M), D)$, $(\Omega(M, T_{poly}), D)$ and $(\Omega(M, \mathcal{D}_{poly}), D)$ have cohomologies concentrated in zero and

$$H^0(\Omega(M, \mathcal{S}M), D) \cong C^\infty(M), \quad (4.46)$$
$$H^0(\Omega(M, T_{poly}), D) \cong \mathcal{F}^0 T_{\text{poly}}, \quad (4.47)$$
$$H^0(\Omega(M, \mathcal{D}_{poly}), D) \cong \mathcal{F}^0 \mathcal{D}_{\text{poly}}. \quad (4.48)$$

Dolgushev constructed an isomorphism μ between $\mathcal{F}^0 T_{\text{poly}}$ and T_{poly} ($\mathcal{F}^0 \mathcal{D}_{\text{poly}}$ and $\mathcal{D}_{\text{poly}}$) and used such a isomorphism to prove that $\Omega(M, T_{\text{poly}})$ is a resolution of T_{poly} (and similarly, $\Omega(M, \mathcal{D}_{\text{poly}})$ is a resolution of $\mathcal{D}_{\text{poly}}$), i.e. the DGLA structure induced on the cohomologies of the complexes $(\Omega(M, T_{poly}), D)$ $((\Omega(M, \mathcal{D}_{poly}), D))$ coincides with the DGLA structure induced from T_{poly} via μ^{-1}. For this reason we refer to $(\Omega(M, T_{poly}), D)$ and $(\Omega(M, \mathcal{D}_{poly}), D)$ as the Fedosov resolutions of T_{poly} and $\mathcal{D}_{\text{poly}}$. These resolutions allow to prove the formality theorem for any smooth manifold. Morally, deforming the quasi-isomorphism (4.35) we get the quasi-isomorphism

$$\underline{U} : (\Omega(M, T_{poly}), D, [\cdot, \cdot]_S) \to (\Omega(M, \mathcal{D}_{poly}), d + D, [\cdot, \cdot]_G). \quad (4.49)$$

This, since $(\Omega(M, T_{poly}), D)$ and $(\Omega(M, \mathcal{D}_{poly}), D)$ are the Fedosov resolutions of T_{poly} and $\mathcal{D}_{\text{poly}}$, gives us the desired quasi-isomorphism

$$U : T_{\text{poly}}(M) \to \mathcal{D}_{\text{poly}}(M). \quad (4.50)$$

4.3 Open Problems

The existence of a star product has been proven by Kontsevich in the case of finite-dimensional manifolds. A natural (but very hard) question involves the case of infinite dimension. This problem has a strong physical motivation, since there are many physical situations where we deal with infinite-dimensional Poisson manifolds. Recently, a discussion on the obstructions for a formality theory in infinite dimension appears in [15, 53].

Another interesting question concerns the quantization of Poisson morphisms; let (M_1, π_1) and (M_2, π_2) be two Poisson manifolds and $\phi : M_1 \to M_2$ a Poisson morphism. The quantization of ϕ should provide a morphism of the associated deformed algebras which gives ϕ in the classical limit. This problem has been approached by Bordemann [6], who related it to symplectic restrictions of star products. However, there is no general solution to this problem yet; the obstructions to the quantization of Poisson morphisms have been showed by Willwacher in [52], where the author gives an explicit counterexample of non quantizable Poisson morphism.

Many questions are related to the different approaches to quantization. Geometric quantization, for instance, is a quantization procedure which focuses on the space of states rather than observables and tries to associate a Hilbert space to a symplectic or Poisson manifold via a complex line bundle. Basically, given a classical phase space M we can define a line bundle L on M. The Hilbert space of square-integrable sections of L is called prequantum Hilbert space H_0. The quantum phase space is a subspace of H_0 and is constructed using global sections of the line bundle which are flat along a polarisation. The reader can find many interesting references in geometric quantization; in particular, besides the historical papers [29, 30] and [26], a comprehensive textbook is the one by Woodhouse [54]. Some introductory treatment can also be found in [3, 32, 36, 42]. The relation between geometric and deformation quantization has been investigated in some specific cases (see for example [21, 22]) but there are still many open questions. So far, the comparison in the case of real polarization, i.e. when there are no global sections which are flat along the polarization, has never been approached.

As already pointed out in Sect. 3.2, a non-formal approach to quantization is given by strict deformation quantization, where the quantum algebra of observables is a C^*-algebra (see [40, 41]). Under some conditions, one can reproduce formal deformation quantization from a C^*-algebraic deformation quantization (an interesting example can be found in [4]). Convergence of formal power series in formal deformation quantization is discussed for instance in [7, 37]. The inverse problem is still open: given a formal deformation quantization we could expect that the subalgebra of the converging power series leads to a strict deformation quantization. However, the only case where this relation is understood is the case of $\mathbb{R}^{2n}$, with the star product introduced in Example 3.1.

A possible future direction of research is the deformation quantization of symplectic groupoids. Symplectic groupoids have been defined to study the quantization of Poisson manifolds (in [48–51]). A symplectic groupoid is a manifold Γ with a multiplication which is only partially defined and compatible with the symplectic structure. The identity elements in Γ form a Poisson manifold M: this correspondence generalizes the one between Lie groups and Lie algebras (this formalism has been largely developed in the last few years; basic introductions can be find in [33–35]). Furthermore, symplectic groupoids could be useful in the study of non-linear commutation relations. The quantization of symplectic groupoids has been studied in the setting of geometric quantization. In particular, the prequantization is discussed in [5, 14, 31]; a notion of polarization of symplectic groupoids, which yields to a strict deformation quantization of the underlying Poisson manifold, has been introduced in [23]. The deformation quantization of symplectic groupoids is an open problem, which could be approached with different techniques, from Fedosov's construction to formality theorem. In this approach, the structures on the symplectic groupoids, a Fedosov connection or a deformed product respectively, can be related with the relative structures on the associated Poisson manifolds.

References

1. I. Batalin, G. Vilkovisky, Gauge algebra and quantization. Phys. Lett. B **102**(1), 27–31 (1981)
2. I. Batalin, G. Vilkovisky, Quantization of gauge theories with linearly dependent generators. Phys. Rev. **D 28**, 2567–2582 (1983)
3. S. Bates, A. Weinstein, *Lectures on the geometry of quantization, Berkeley Mathematics Lecture Notes*. vol. 8, AMS, Providence, (1997)
4. P. Bieliavsky, V. Gayral, Deformation Quantization for Actions of Kählerian Lie Groups. Mem. Am. Math. Soc. (to appear) (2011)
5. F. Bonechi, A. Cattaneo, M. Zabzine, Geometric quantization and non-perturbative Poisson sigma model. Adv. Theor. Math. Phys. **10**(5), 683–712 (2006)
6. M. Bordemann, The deformation quantization of certain super-Poisson brackets and BRST cohomology. ed. by G. Dito, D. Sternheimer, in *Conference Moshé Flato* (1999)
7. M. Bordemann, M. Brischle, C. Emmrich, S. Waldmann, Subalgebras with Converging Star Products in Deformation Quantization: An Algebraic Construction for $\mathbb{C}P^n$. J. Math. Phys. **37**(12), 6311–6323 (1996)
8. S. Bosch, *Lectures of Formal and Rigid Geometry*, Lecture Notes in Mathematics (Springer, Berlin, 2014)
9. A. Cattaneo, G. Felder, A path integral approach to the Kontsevich quantization formula. Comm. Math. Phys. **212**, 591–611 (2000)
10. A.S. Cattaneo, G. Felder, L. Tomassini, From local to global deformation quantization of Poisson manifolds. Duke Math. J., **115**, 329–352 (2002)
11. A.S. Cattaneo, G. Felder, On the globalization of Kontsevich's star product and the perturbative Poisson sigma model. Prog. Theor. Phys. Suppl. **144**, 38–53 (2001)
12. A.S. Cattaneo, G. Felder, L. Tomassini, *Deformation Quantization, IRMA Lectures in Mathematics and Theoretical Physics*. Fedosov connections on jet bundles and deformation quantization, (2002)
13. A.S. Cattaneo, Formality and star products. *Poisson Geometry, Deformation Quantization and Group Representations*, vol. 323 in London Mathematical Society Lecture Note series, (Cambridge University Press, 2005), pp. 79–144
14. M. Crainic, Prequantization and Lie brackets. J. Symplectic. Geom., **2**(4), (2004)
15. G. Dito, The necessity of wheels in universal quantization formulas. arXiv:1308.4386
16. V.A. Dolgushev, Covariant and equivariant formality theorems. Adv. Math. **191**(1), 147–177 (2005)
17. V.A. Dolgushev, A proof of Tsygan's formality conjecture for an arbitrary smooth manifold. Ph.D thesis, MIT (2005)
18. P. Etingov, MIT Lecture notes. http://ocw.mit.edu/courses/mathematics/18-238-geometry-and-quantum-field-theory-fall-2002/lecture-notes/
19. B.V. Fedosov, A simple geometrical construction of deformation quantization. J. Differ. Geom. **40**, 213–238 (1994)
20. I.M. Gelfand, Some problems of differential geometry and the calculation of the cohomology of Lie algebras of vector fields. Sov. Math. Dokl. **12**, 1367–1370 (1971)

21. E. Hawkins, The correspondence between geometric quantization and formal deformation quantization. arXiv:math/9811049 (1998)
22. E. Hawkins, Geometric quantization of vector bundles and the correspondence with deformation quantization. Comm. Math. Phys. **215**(2), 409–432 (2000)
23. E. Hawkins, A groupoid approach to quantization. J. Symplectic Geom. **6**(1), 61–125 (2008)
24. V. Hinich, Tamarkin's proof of Kontsevich's formality theorem. Forum Math. **15**, 591–614 (2003)
25. B. Keller, *Deformation quantization after Kontsevich and Tamarkin*. In: Déformation, quantification, théorie de Lie, vol. 20 (Panoramas et Synthèses, Société mathématique de France, 2005), pp. 19–62
26. A.A. Kirillov, *Geometric quantization. In Dynamical Systems IV*, Encyclopaedia of Mathematical Sciences. (Springer, Berlin, 1990)
27. I. Kolar, P.W. Michor, J. Slovak, *Natural operations in differential geometry* (Springer, 2010)
28. M. Kontsevich, Deformation quantization of Poisson manifolds. Lett. Math. Phys. **66**, 157–216 (2003)
29. B. Kostant, *Quantization and unitary representation. In Lectures in Modern Analysis and Applications III*, vol. 170 of Lecture Notes in Mathematics (Springer, Berlin, 1970), pp. 87–208
30. B. Kostant, On the definition of quantization. In: Géométrie Symplectique et Physique Mathématique, vol. 237 of Colloques Intern. CNRS, (1975)
31. C. Laurent-Gengoux, P. Xu, Quantization of pre-quasi-symplectic groupoids and their hamiltonian spaces. In: ed. by J.E. Marsden, T.S. Ratiu, The Breadth of Symplectic and Poisson Geometry: Festschrift in Honor of Alan Weinstein, vol. 232 of Progress in Mathematics. Birkhäuser, (2005)
32. E. Lerman, Geometric quantization: a crash course. Contemp. Math. **583**, 147–174 (2012)
33. K.C.H. Mackenzie, *Lie Groupoids and Lie Algebroids in Differential Geometry*, Number 124 in London Mathamatical Society Lecture notes series (Cambridge University Press, Cambridge, 1987)
34. K.C.H. Mackenzie, *General Theory of Lie Groupoids and Lie Algebroids*, London Math. Soc. Lecture notes series (Cambridge University Press, Cambridge, 2005)
35. C.M. Marle, *Lie, Symplectic and Poisson Groupoids and Their Lie Algebroids*, Encyclopedia of Mathematical Physics (Elsevier, Amsterdam, 2006)
36. E. Miranda, *From action-angle coordinates to geometric quantization and back*, In: Finite Dimensional Integrable Systems in Geometry and Mathematical, 2011, (2011)
37. H. Omori, Y. Maeda, N. Miyazaki, A. Yoshida, Deformation quantization of Fréchet-Poisson algebras of Heisenberg type. Contemp. Math. **288**, 391–395 (2001)
38. M.E. Peskin, D.V. Schroeder, *An Introduction to Quantum Field Theory* (Westview Press, Boulder, 1995)
39. M. Polyak, Quantization of linear Poisson structures and degrees of maps. Lett. Math. Phys. **66**(1–2), 15–35 (2003)
40. M. Rieffel, Deformation quantization and operator algebra. *Proceedings of Symposia in Pure Mathematics*, vol. 51, (1990)
41. M. Rieffel, Quantization and C^*-algebras. Contemp. Math., **167**, (1994)
42. W.G. Ritter, Geometric quantization. arXiv:math-ph/0208008
43. L.H. Ryder, *Quantum Field Theory* (Cambridge University Press, Cambridge, 1985)
44. D.J. Saunders, *The Geometry of Jet Bundles*, Number 142 in London Math. Soc. Lecture notes series (Cambridge University Press, Cambridge, 1989)
45. B. Shoikhet, Vanishing of the Kontsevich Integrals of the Wheels. Lett. Math. Phys. **56**, 141–149 (2001)
46. D.E. Tamarkin, Another proof of M. Kontsevich formality theorem. arXiv:math/9803025, (1998)
47. D.E. Tamarkin, Formality of chain operad of little discs. Lett. Math. Phys. **66**(1–2), 65–72 (2003)

48. A. Weinstein, Symplectic groupoids and Poisson manifolds. Bull. Amer. Math. Soc. **16**(1), 101–104 (1987)
49. A. Weinstein, *Symplectic groupoids, geometric quantization, and irrational rotation algebras*, In Symplectic Geometry, Groupoids and Integrable Systems, vol. 20 of Mathematical Sciences Research Institute Publications (Springer, Berlin, 1991)
50. A. Weinstein, *Tangential deformation quantization and polarized symplectic groupoids*, In: Deformation Theory and Symplectic Geometry, vol. 20 of Mathematical Physics Studies (Springer, Berlin, 1996)
51. A. Weinstein, P. Xu, Extensions of symplectic groupoids and quantization. J. Reine Angew. Math. **417**, 159–189 (1991)
52. T. Willwacher, Counterexample to the quantizability of modules. Lett. Math. Phys. **81**(3), 265–280 (2007)
53. T. Willwacher, The obstruction to the existence of a loopless star product. arXiv:1309.7921, (2013)
54. N.M.J. Woodhouse, *Geometric Quantization*, 2nd edn. (Oxford University Press, Oxford, 1997)

Appendix A

This Appendix aims to be a short survey of some basic notions used in this book; in particular, complexes and cohomologies, vector bundles and connections on vector bundles. The reader is assumed to be familiar with basic notions of differential geometry such as smooth manifolds, vector fields and differential forms (see e.g., [1, 3, 4]).

A.1 Vector Bundles

Definition A.1 A vector bundle of rank m consists of a pair of manifolds E and M, with a (smooth) surjective map $\pi : E \to M$ such that, for any $x \in M$, the subset $E_x = \pi^{-1}(x) \subset E$ is a vector space isomorphic to k^m[1], and there exists an open neighborhood U of x in M and a so-called local trivialization

$$\boldsymbol{\Phi} : \pi^{-1}(U) \to U \times k^m. \tag{A.1}$$

$\boldsymbol{\Phi}$ is a diffeomorphism which restricts to a linear isomorphism $E_y \to \{y\} \times k^m$ for any $y \in U$. We call E the total space of the bundle $\pi : E \to M$ and M the base. For any $x \in M$, the set $E_x = \pi^{-1}(x) \subset E$ is called the fiber over x.

Given a vector bundle $\pi : E \to M$, we denote $E|_U = \pi^{-1}(U)$ for any subset $U \subset M$. We say that the bundle is trivializable over the subset U if there exists a trivialization $\boldsymbol{\Phi} : E|_U \to U \times k^m$. The bundle is said to be globally trivializable (or trivial) if there exists a trivialization over the entire manifold M. Every vector bundle admits a system of local trivializations, i.e. a covering of M by open sets U_i and diffeomorphisms $\boldsymbol{\Phi}_i : E|_{U_i} \to U_i \times k^m$. Such a system defines a set of continuous transition maps

---

[1] Here k can be either $\mathbb{R}$ or $\mathbb{C}$

© The Author(s) 2015
C. Esposito, *Formality Theory*, SpringerBriefs in Mathematical Physics,
DOI 10.1007/978-3-319-09290-4

$$f_{ij} : U_i \cap U_j \to GL(m, k), \qquad (A.2)$$

so that the diffeomorphism $\Phi_i \cdot \Phi_j^{-1} : (U_i \cap U_j) \times k^m \to (U_i \cap U_j) \times k^m$ has a form $\Phi_i \cdot \Phi_j^{-1} : (x, u) \mapsto (x, f_{ij}(x)u)$.

Definition A.2 A section of the bundle $\pi : E \to M$ is a map $u : M \to E$ such that $(\pi \cdot u)(x) = x$ for every $x \in M$.

The space of (smooth) sections on a vector bundle $\pi : E \to M$ is denoted by

$$\Gamma(E) = \{u : M \to E | \pi \cdot u = Id_M\} \qquad (A.3)$$

Example A.1 Given a n-dimensional smooth manifold M, its tangent bundle $TM = \bigcup_{x \in M} T_x M$ associates to each $x \in M$ the n-dimensional vector space $T_x M$. In this case, M is the base of the bundle, the $2n$-dimensional manifold TM is the total space. There a natural projection map $\pi : TM \to M$ which, for any $x \in M$, maps every vector $X \in T_x M$ to x. The fibers are given by the preimages $\pi^{-1}(x) = T_x M$.

Example A.2 The dual version of the tangent bundle is called cotangent bundle and is denoted by $T^*M \to M$. Its fibers are the vector spaces $T_x^* M$ of linear maps $T_x M \to \mathbb{R}$, called dual vectors. The sections of T^*M are the differential 1-forms on M.

A.1.1 Tensors

Consider the set $L^k(V_1, \ldots, V_k; W)$ of k-multilinear maps of $V_1 \times \cdots V_k$ to W. The special case $L(V, \mathbb{R})$ is denoted V^*, the dual space of V. If V is finite dimensional and $\{e_1, \ldots e_n\}$ is a basis of V, there is a unique basis of V^*, the dual basis $\{f^1, \ldots f^n\}$, such that $\langle f^i, e_j \rangle = \delta_j^i$. Here $\langle \cdot, \cdot \rangle$ denotes the pairing between V and V^*.

For a vector space V we put

$$T_s^r(V) = L^{s+r}(V^*, \ldots, V^*, V, \ldots, V; \mathbb{R}) \qquad (A.4)$$

(r copies of V^* and s copies of V). Elements of $T_s^r(V)$ are called tensors on V, contravariant of order r and covariant of order s.

Given $t \in T_s^r(V)$ and $s \in T_p^q(V)$, the tensor product of t and s is the tensor $t \otimes s \in T_{s+p}^{r+q}(V)$ defined by

$$(t \otimes s) \ (\beta^1, \ldots \beta^r, \gamma^1, \ldots, \gamma^q, f_1, \ldots f_s, g_1, \ldots, g_p)$$
$$= t(\beta^1, \ldots \beta^r, f_1, \ldots f_s)s(\gamma^1, \ldots, \gamma^q, g_1, \ldots, g_p) \qquad (A.5)$$

where $\beta^j, \gamma^j \in V^*$ and $f_j, g_j \in V$.

The tensor product is associative, bilinear and continuous; it is not commutative. Notice that

$$T_0^1(V) = V, \quad T_1^0(V) = V^*. \tag{A.6}$$

Let M be a manifold and $\pi : TM \to M$ its tangent bundle. We call $T_s^r(M) = T_s^r(TM)$ the vector bundle of tensors contravariant of order r and covariant of order s. We identify $T_0^1(M)$ with TM and call $T_1^0(M)$ the cotangent bundle of M also denoted by $\tau_M^* : T^*M \to M$. The zero section of $T_s^r(M)$ is identified with M.

A section of $T_s^r(M)$ takes an element $m \in M$ and associates a vector in the fiber, called tensor.

A tensor field of type (r, s) on a manifold M is a smooth section of $T_s^r(M)$. We denote by $\mathscr{T}_s^r(M)$ the set $\Gamma(T_s^r(M))$ together with its infinite dimensional real vector space structure. A covector field or differential one-form is an element of $\mathscr{T}_1^0(M)$.

A.1.2 Connections

Let us consider a rank n vector bundle E over a m-dimensional manifold M and the cotangent bundle T^*M. The tensor product $E \otimes T^*M$ is obviously a vector bundle over M, with fibers $(E \otimes T^*M)_x = E_x \otimes T_x^*M$.

Definition A.3 A connection D in a vector bundle E is a first-order differential operator $D : \Gamma(E) \to \Gamma(E \otimes T^*M)$ satisfying the Leibniz rule

$$D(fu) = (\mathrm{d}f)u + fDu \tag{A.7}$$

for any section $u \in \Gamma(E)$ and any function $f \in C^\infty(M)$; here $\mathrm{d}f$ is the differential of f.

The section $Du \in \Gamma(E \otimes T^*M)$ is called covariant differential of u. Let $e_U = (e^1, \dots, e^n)_U$ be local coordinates of E, where U is an open set $U \subset M$; the definition of connection is local, thus we only need to define the operator D using the local frame e_U. In particular, we can write $De_j = \Gamma_j^k e_k$, with some one-forms Γ_j^k defined on U. The matrix $\Gamma_U = (\Gamma_j^k)_U$ is called local connection one-form.

Proposition A.1 *There exists a connection for any vector bundle.*

In the following, we denote by Ω^p the p-th exterior power of T^*M and $\Omega = \bigoplus_{p=0}^m \Omega^p$; consider the square of the operator D

$$D^2 = D \cdot D : \Gamma(E) \to \Gamma(E \otimes \Omega^2). \tag{A.8}$$

Then D^2 is a tensor and we set, for any section $u \in \Gamma(E)$,

$$D^2 u = F u; \tag{A.9}$$

F is called curvature of the connection D.

Definition A.4 A connection D is said to be flat if $F = 0$.

Consider a smooth manifold M with local coordinates $(x^1, \ldots, x^n)$ and its tangent bundle TM. The vector fields $\frac{\partial}{\partial x^1}, \ldots, \frac{\partial}{\partial x^n}$ and the differentials $dx^1, \ldots, dx^n$ form a local system of coordinates of TM and T^*M respectively, called natural. Like any other bundle, the tangent bundle can be endowed by a connection, called affine connection.

Definition A.5 A fundamental one-form θ on M is a section of $\Omega \otimes TM$ such that the $i(X)\theta = X$ for any X, where $i(X)$ is the contraction of the vector field X with differential forms. If D is an affine connection, the torsion of D is the differential form $S = D\theta \in \Gamma(\Omega^2 \otimes TM)$.

If $S = 0$, the connection D is said to be torsion free.

In local coordinates, we have $\theta = e_i \theta^i$ or $\theta = dx^i \frac{\partial}{\partial x^i}$ (natural coordinates) and

$$S = D(e_i \theta^i) = De_i \wedge \theta^i + e_i d\theta^i = e_i(\Gamma_j^i \wedge \theta^j + d\theta^i) = \Gamma_i^j \wedge dx^i \frac{\partial}{\partial x^j}. \tag{A.10}$$

Since $\Gamma_i^j = \Gamma_{ik}^j dx^k$, we have

$$S = \Gamma_{ik}^j dx^k \wedge dx^i \frac{\partial}{\partial x^j}, \tag{A.11}$$

thus the torsion tensor in natural coordinates is given by

$$S_{ik}^j = \frac{1}{2}(\Gamma_{ki}^j - \Gamma_{ik}^j). \tag{A.12}$$

The coefficients Γ_{jk}^i of the affine connection in natural coordinates are called Christoffel symbols.

A.2 Cohomology

We recall here some basic definitions and results on complexes and cohomologies; the interested reader is referred to the classical literature, e.g. [2].

Definition A.6 A cochain complex $C^\bullet$ is a sequence of vector spaces (more generally, abelian groups) $\{C^n\}_{n \in \mathbb{Z}}$ and homomorphisms $d_n : C^n \to C^{n+1}$ such that, for all n, $d_{n+1} \circ d_n = 0$. The maps d_i are called coboundary operators and the elements in C^n are called n-cochains.

Such complexes are generally represented as a sequence of linear maps

$$\cdots \longrightarrow C^{n-1} \xrightarrow{d_{n-1}} C^n \xrightarrow{d_n} C^{n+1} \longrightarrow \cdots \qquad (A.13)$$

the composition of any two being zero. The elements in $Z^n(C) := Ker(d^n)$, i.e. such that $d_n c^n = 0$ are called n-cocycles and elements in $B^n(C) := Im(d^{n-1})$ are called n-coboundaries. The condition $d_{n+1} \circ d_n = 0$ implies that $B^n(C) \subset Z^{n+1}(C)$ for any $n \in \mathbb{Z}$. Thus the quotient group $\frac{Z^{n+1}}{B^n}$ is defined for all n.

Definition A.7 Let $C^\bullet = \bigoplus_{n \in \mathbb{Z}} C^n$ be a cochain complex of groups. The n-th cohomology group of $C^\bullet$, $H^n(C)$, is defined by

$$H^n(C) = \frac{Z^n(C)}{B^{n-1}(C)}. \qquad (A.14)$$

Elements of $H^n(C)$ are equivalent classes of cocycles: two cocycles are equivalent or cohomologous if their difference is a coboundary. The cohomology of C is the direct sum of vector spaces $H(C) = \bigoplus_{n \in \mathbb{Z}} H^n(C)$.

Definition A.8 A map $f : A \to B$ between two complexes is a chain map if it commutes with the differential operators of A and B

$$f d_A = d_B f. \qquad (A.15)$$

A sequence of vector spaces

$$\cdots \longrightarrow V^{n-1} \xrightarrow{f_{n-1}} V^n \xrightarrow{f_n} V^{n+1} \longrightarrow \cdots \qquad (A.16)$$

is said to be exact if for all n the kernel of f_n is equal to the image of its predecessor f_{n-1}. An exact sequence of the form

$$0 \to A \to B \to C \to 0 \qquad (A.17)$$

is called a short exact sequence.

Example A.3 (de Rham cohomology) The best known example is the so-called de Rham complex, which is the cochain complex of exterior differential forms on a smooth manifold M, endowed with the exterior derivative. A nice introduction of the de Rham cohomology can be found in [5]. Let M an open set in $\mathbb{R}^n$ and $x_1, \ldots, x_n$ be linear coordinates on $\mathbb{R}^n$. We define $\Omega^\bullet$ to be the algebra over $\mathbb{R}$ generated by $dx_1, \ldots, dx_n$ with

$$\begin{cases} (dx_i)^2 = 0, \\ dx_i dx_j = -dx_j dx_i. \end{cases} \qquad (A.18)$$

The differential forms on $\mathbb{R}^n$ are elements of $\Omega^\bullet(\mathbb{R}^n) = C^\infty(\mathbb{R}^n) \otimes \Omega^\bullet$. Thus, a form α can be uniquely written as $\sum f_{i_1...i_q} dx_{i_1} \ldots dx_{i_q} = \sum f_I dx_I$, where the coefficients $f_{i_1...i_q}$ are smooth functions on $\mathbb{R}^n$. The algebra $\Omega^\bullet(\mathbb{R}^n) = \bigoplus_{q=0}^{n} \Omega^q(\mathbb{R}^n)$ is graded and $\Omega^q(\mathbb{R}^n)$ consists of the smooth q-forms on $\mathbb{R}^n$.

We can define a differential operator

$$d : \Omega^q(\mathbb{R}^n) \to \Omega^{q+1}(\mathbb{R}^n),$$ (A.19)

by

1. if $f \in \Omega^0(\mathbb{R}^n)$, then $df = \sum \frac{\partial f}{\partial x_i} dx_i$,
2. if $\alpha = \sum f_I dx_I$, then $d\alpha = \sum df_I dx_I$.

This d is called exterior differentiation.

The wedge product of two differential forms $\alpha = \sum f_I dx_I$ and $\beta = \sum g_J dx_J$ is given by

$$\alpha \wedge \beta = \sum f_I g_J dx_I dx_J.$$ (A.20)

It can be proven (see [2]) that d is an anti-derivation, i.e.

$$d(\alpha \wedge \beta) = (d\alpha) \wedge \beta + (-1)^{\deg \alpha} \alpha \wedge d\beta.$$ (A.21)

Finally, it is easy to check that $d^2 = 0$. The complex $\Omega^\bullet(\mathbb{R}^n)$ with the differential d is called de Rham complex on $\mathbb{R}^n$. The kernel of d are the closed forms and the image of d the exact forms. The qth de Rham cohomology of $\mathbb{R}^n$ is the vector space

$$H_{dR}^q(\mathbb{R}^n) = \{\text{closed } q - \text{forms}\}/\{\text{exact } q - \text{forms}\}.$$ (A.22)

Example A.4 Consider the groups

$$C^0 = \{f \in C^\infty(\mathbb{R}^2)\}, \quad C^1 = \{f dx + g dy : f, g \in C^0\}, \quad C^2 = \{f dx dy : f \in C^0\},$$ (A.23)

with

$$d^0 : f \mapsto f_x dx + f_y dy, \quad d^1 : f dx + g dy \mapsto (g_x - f_y) dx dy.$$ (A.24)

We observe that

$$d^1 \circ d^0 : f \mapsto f_x dx + f_y dy \mapsto (f_{yx} - f_{xy}) dx dy = 0;$$ (A.25)

thus, defining $C^i = 0$ and $d^i = 0$ for all $i < 0$ and $i > 2$, we have the cochain complex $C^\bullet$:

$$\cdots 0 \to 0 \xrightarrow{\iota} C^0 \xrightarrow{d^0} C^1 \xrightarrow{d^1} C^2 \xrightarrow{0} 0 \to 0 \to \cdots$$ (A.26)

Example A.5 (Chevalley-Eilenberg cohomology) Let $\mathfrak{g}$ be a finite-dimensional Lie algebra and ρ be a representation $\rho : \mathfrak{g} \to End(V)$:

$$\rho(X)\rho(Y) - \rho(Y)\rho(X) = \rho([X, Y]) \qquad (A.27)$$

for all $X, Y \in \mathfrak{g}$. Define the space of linear maps

$$C^n(\mathfrak{g}, V) := Hom(\wedge^n \mathfrak{g}, V) \cong \wedge^n \mathfrak{g}^* \otimes V \qquad (A.28)$$

called the space of n-forms on $\mathfrak{g}$ with values in V. This space can be endowed with a differential $d : C^n(\mathfrak{g}, V) \to C^{n+1}(\mathfrak{g}, V)$ as follows:

1. for $v \in V$, $dv(X) = \rho(X)v$ for all $X \in \mathfrak{g}$;
2. for $\alpha \in \mathfrak{g}^*$, let $d\alpha(X, Y) = -\alpha([X, Y])$ for all $X, Y \in \mathfrak{g}$;
3. for $\omega \otimes v \in \wedge^\bullet \mathfrak{g}^* \otimes V$, $d(\omega \otimes v) = d\omega \otimes v + (-1)^{|\omega|} \omega \wedge dv$.

It can be checked that $d^2 = 0$ everywhere. Thus, we have a complex

$$\cdots \longrightarrow C^{n-1}(\mathfrak{g}, V) \xrightarrow{d} C^n(\mathfrak{g}, V) \xrightarrow{d} C^{n+1}(\mathfrak{g}, V) \longrightarrow \cdots \qquad (A.29)$$

called Chevalley–Eilenberg complex of $\mathfrak{g}$ with values in V. Its cohomology is called Lie algebra cohomology of $\mathfrak{g}$ with values in V.

References

1. R. Abraham, J.E. Marsden, T. Ratiu, *Manifolds, Tensor Analysis, and Applications* (Springer, New York, 1988)
2. R. Bott, L.W. Tu, *Differential Forms in Algebraic Topology* (Springer, New York, 1982)
3. S.S. Chern, W.H. Chen, K.S. Lam, Lectures on Differential Geometry. *Series on University Mathematics* (World Scientific Pub. Co., Inc., Singapore, 1999).
4. J.M. Lee, *Introduction to Smooth Manifolds* (Springer, New York, 2002)
5. I. Madsen, J. Tornehave, *From Calculus to Cohomology: De Rham cohomology and characteristic classes* (Cambridge University Press, Cambridge, 1997)

Index

© The Author(s) 2015
C. Esposito, *Formality Theory*, SpringerBriefs in Mathematical Physics,
DOI 10.1007/978-3-319-09290-4